Topics in Applied Physics Volume 17

Topics in Applied Physics   Founded by Helmut K. V. Lotsch

Electroluminescence

Edited by J. I. Pankove

With Contributions by
P. J. Dean   T. Inoguchi   S. Mito   J. I. Pankove
Y. S. Park   B. K. Shin   Y. M. Tairov   Y. A. Vodakov
S. Wagner

With 127 Figures

Springer-Verlag Berlin Heidelberg New York 1977

Dr. *Jacques I. Pankove*

RCA Laboratories, David Sarnoff Research Center
Princeton, NJ 08540, USA

ISBN 3-540-08127-5 Springer-Verlag Berlin Heidelberg New York
ISBN 0-378-08127-5 Springer-Verlag New York Heidelberg Berlin

Library of Congress Cataloging in Publication Data. Main entry under title: Electroluminescence. (Topics in applied physics; v. 17) Includes bibliographical references and index. 1. Electroluminescence. I. Pankove, Jacques I., 1922−. II. Dean, Paul Jeremy, 1935−. QC480.E43   535′.35   77−1911

© by Springer-Verlag Berlin Heidelberg 1977

Printed in Germany

Monophoto typesetting, offset printing, and bookbinding: Brühlsche Universitätsdruckerei, Lahn-Giessen

2153/3130−543210

Preface

Although electroluminescence has been known for more than half a century, its utilization has become practical only within the last decade. The major attributes of EL devices are compactness, ruggedness, and long operational life. But, much progress remains to be achieved mostly with respect to efficiency and cost, and, for large-area displays, compatibility with solid-state integrated circuits.

It is hoped that this book will serve two purposes:

1) to educate newcomers to this exciting area of physics and technology, providing in quick brush strokes a background of fundamentals highlighted with reviews of recent technological developments;

2) to provide specialists with useful references and new insights in adjoining areas of luminescence.

The contributors were selected for their long and continuing expertise in the study of luminescence in selected compounds. Since the boundaries of present knowledge have been outlined by each author, this volume should serve as a stepping stone for future progress.

Princeton, New Jersey   *J. I. Pankove*
November 1976

Contents

Contributors

Dean, Paul J.
: Royal Signals and Radar Establishment, St. Andrews Road, Gt. Malvern, Worcestershire, England

Inoguchi, Toshio
Mito, Sanai
: Central Research Laboratories, Engineering Division, Sharp Corporation, Tenri, Nara 632, Japan

Pankove, Jacques I.
: RCA Laboratories, David Sarnoff Research Center, Princeton, NJ 08540, USA

Park, Yoon Soo
: Air Force Avionics Laboratory, Wright-Patterson Air Force Base, Dayton, OH 45433, USA

Shin, B. K.
: Systems Research Laboratories, Inc., Dayton, OH 45440, USA

Tairov, Yuri M.
: V. I. Ulyanov-Lenin, Leningrad Institute of Electrical Engineering, 197022, Leningrad, USSR

Vodakov, Yuri A.
: A. F. IOFFE, Institute of Technical Physics, Academy of Sciences of the USSR, 194021, Leningrad, USSR

Wagner, Sigurd
: Bell Telephone Laboratories, Holmdel, NJ 07733, USA

1. Introduction

J. I. Pankove

With 24 Figures

In electroluminescence, electrical energy is transformed into light. Therefore, two processes are of interest 1) how the radiative system is excited, and 2) the light generation mechanism itself. We shall first review the light generation mechanism whereby various transitions are made from upper to empty lower energy states. Although only the radiative transitions are useful, we must also consider the competing non-radiative processes which adversely affect the luminescence efficiency. We shall next survey the various modes of electrical excitation employed in electroluminescent devices. Finally, we shall consider psychophysiological factors which guide our choice in material selection.

1.1 Radiative Transitions

A photon can be emitted when an electron drops from an upper to a lower energy level. These levels can be either intrinsic band states or impurity levels.

1.1.1 Band-to-Band Transitions

These transitions, involving free holes and free electrons, are sometimes called free-to-free transitions. We shall not discuss free exciton recombination because our emphasis will be on those room-temperature processes which are likely to be involved in electroluminescent devices. Direct-gap materials (all the II-VI and most III-V compounds) permit momentum-conserving transitions between conduction and valence bands (Fig. 1.1). Indirect-gap materials (group IV semiconductors and some III-V compounds) require the emission or absorption of a phonon to complete the lowest energy transition across the energy gap. Band-to-band radiative transitions are therefore much more probable across a direct than across an indirect gap.

Note that if an optical transition between two states is allowed, it can proceed in either direction: emission or absorption. It is for this reason that transitions across the energy gap have low practical interest; the band-to-band radiation in a direct gap material can be readily reabsorbed by exciting a valence band electron into the conduction band.

Phonon emission can shift the band-to-band spectrum to lower energies by multiples of the phonon energy. Since this is a multistep process, its probability is lower than for the direct recombination. However, the phonon-

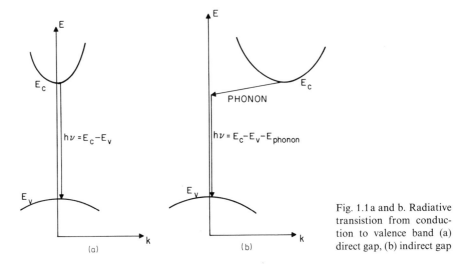

Fig. 1.1a and b. Radiative transistion from conduction to valence band (a) direct gap, (b) indirect gap

shifted lower energy emission may be more readily transmitted by the material due to its reduced absorption at lower than bandgap energies.

1.1.2 Transitions Between Band and Impurity

A radiative transition can occur between an impurity state and an intrinsic band (conduction-band-to-acceptor or donor-to-valence-band). Such a free-to-bound transition can be highly probable if the impurity is deep. The wave function of deep levels extends throughout the momentum space, thus permitting momentum-conserving transitions even in indirect gap materials.

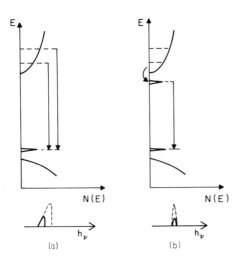

Fig. 1.2a and b. Effect of band-filling on the emission spectrum to determine whether band states are involved in the radiative transition (a) or not (b)

Techniques for determining the depths of impurity levels and for characterizing their properties are discussed in Chapter 3.

A simple method to find whether or not a band is involved in a radiative transition consists of increasing the population of that band. If a band is involved, the emission spectrum should shift to higher energies as shown in Fig. 1.2a. The population of the band can be increased either by heavier doping or by higher excitation levels. However, one must be cautious in the case of injection electroluminescence, since the applied voltage not only controls the population via the quasi-Fermi levels, but also moves the relative position of the various levels (for example, those of the p-type side with respect to those of the n-type region). In such questionable cases, it is best to use optical excitation (photoluminescence) to control the population of the band. When band states are not involved in the recombination process (Fig. 1.2b), the emission intensity increases with excitation without a change in the spectrum.

1.1.3 Transitions at a Localized Center

In wide gap materials, it is possible to excite an impurity without ionizing it, i.e., without placing its electron in the conduction band. Radiative transitions from such localized states are observed in many phosphors (e.g., ZnS : Mn). When the electron of such a deep center returns to the ground state, it emits a narrow spectrum characteristic of an atomic transition. It is as though one were dealing with a gas of radiating atoms dispersed in a medium having a higher dielectric constant than if these atoms formed a true gas outside the host crystal. Further complications may result from a possible anisotropy of the ambient.

1.1.4 Transitions at Isoelectronic Centers

Isoelectronic centers are formed by replacing one atom of the host crystal by another atom of the same valence, e.g., N replacing P in GaP. An isoelectronic center can capture an electron in a short range potential thus becoming charged and very attractive to a hole; then, a bound exciton is formed [1.1]. Because the exciton bound to an isoelectronic center (or at least one of the carriers) is very localized (wave function very spread in momentum space), it has a high probability of recombining radiatively. When the concentration of these centers, say N in GaP, is increased, the bound excitons interact, increasing their binding energy, thus lowering the energy of the excited state and of the emitted photon. Nearest pairs are most tightly bound and emit the lowest photon energy.

1.1.5 Donor-to-Acceptor Transitions

Transitions between donors and acceptors can be very efficient. Most light-emitting diodes use this mechanism [1.2, 3]. In the phosphor literature, accep-

tors and donors are called "activators and coactivators" [1.4]. Such pairs can
be thought of as molecular impurities when they occur on neighboring sites.
The Coulomb interaction between a donor and an acceptor is a strong function
of the overlap of the electron and hole wave functions. The Coulomb interaction
increases the energy of the excited state by an increment which is inversely
proportional to the separation r between the donor and the acceptor. The
energy of the photon resulting from a donor-acceptor transition is given by
[1.2, 5]:

$$hv = E_g - (E_D + E_A) + q^2/\varepsilon r \tag{1.1}$$

where E_D and E_A are the donor and acceptor binding energies, respectively,
and ε is the dielectric constant of the host crystal. Fig. 1.3 illustrates this
mechanism.

In some materials (e.g., Ge, GaAs), the Coulomb energy increment for the
nearest pairs can be larger than the activation energies of the impurities
$E_D + E_A$, thus making the excitation energy for that pair greater than the
energy gap [1.6]. Once excited, such an electron-hole pair will relax to the band
edges and recombine via some other transition, such as via a more distant pair.

The emission spectrum of donor-to-acceptor transitions is broad, owing
to the large range of separations possible between impurities. The nearest
pairs contribute the highest energy photons, often in well-resolved spectral
peaks. Distant pairs contribute the lowest energy photons in a rather broad
band.

Note that the farther apart the donor and acceptor are, the farther an
electron must tunnel to complete a radiative transition to the acceptor. In
other words, the tunneling probability decreases with increasing distance. A
low probability transition is a slow process. A distinguishing characteristic of

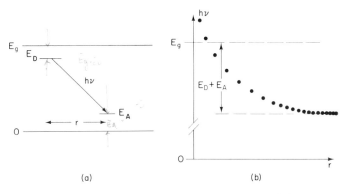

Fig. 1.3a and b. Donor-to-acceptor transition (a) effect of Coulomb interaction on emission
energy, (b) r is the donor-acceptor separation

donor-to-acceptor transitions is their time-resolved spectra: if one observes the decay of luminescence after the excitation is turned off, one finds that the emission spectrum gradually shifts to lower energies. The nearest pairs contribute the highest energies with a fast decay rate, while the distant pairs produce the low energy photons with the slowest decay [1.7].

Relatively deep donors and acceptors are desirable for an efficient radiative recombination. This is due to the following factors: 1) momentum conservation is easier with deep impurities, as mentioned in Subsection 1.1.2; 2) the binding energy of deep nearest pairs can be lower than the energy gap; 3) recombination at nearest pairs does not require tunneling.

At very low temperatures, sharp spectral lines can be resolved and assigned to specific pair separations. For an accurate interpretation, one must take into account the fact that the overlap of electron and hole wave functions is modified by local strains generated by the substitutional insertion of a differently sized atom [1.8].

Since the impurities are located subsitutionally at lattice sites, various orders of pairing fall on discrete shells. Note that the spacing between the discrete shells becomes infinitesimally small for the most distant pairs; this merging of the shells results in a broad emission spectrum. In the resolvable range, however, the spectrum permits one to distinguish whether the donors and acceptors are substituting at similar sites (e.g., Sn and Te on P-sites in GaP) or at different sites (e.g., Zn at Ga-site and O at P-site).

The red emission of GaP doped with Zn and O has been shown to result from a different mechanism [1.9, 10]: The Zn which substitutes on a Ga-site is an acceptor while O on a P-site is a donor. When Zn and O occupy adjacent sites (nearest pair) they form a neutral molecular center. Just like an isoelectronic impurity, this center can trap an electron-hole pair to form a bound exciton [1.9]. Alternatively, such a center can trap an electron, subsequently behaving as a donor; a nearby isolated Zn can trap a hole which then interacts with the electron on the Zn-O electron trap to form another type of bound exciton [1.10]. The recombination of these excitons emits red light. These are the dominant mechanisms at room temperature for GaP : Zn, O LEDs. After recombination, the Zn-O molecular center is neutral, therefore there is no Coulomb interaction between this donor-like complex and the isolated Zn acceptor.

1.2 Non-Radiative Recombination

When a transition from an upper to a lower energy state can occur without emitting a photon, the emission efficiency is decreased. Non-radiative recombination processes are difficult to identify or to study directly because their occurrence can only be inferred from the low emission efficiency. Several mechanisms are possible.

1.2.1 Multiphonon Emission

Multiphonon emission is best appreciated by looking at a configuration co-ordinate diagram (Fig. 1.4), which shows the energy of an electron in both the ground state (lower curve) and in the excited state (upper curve) as a function of a generalized coordinate representing the average distance of atoms surrounding the atom with which this electron is associated. If the atom moves to r_c (e.g., due to thermal vibration) the electron can escape the excited state at the c crossing of the two levels and return to the minimum energy at r_0 while the atoms move to the corresponding lower equilibrium distance. The atomic displacement generates phonons. Since the phonon energy is much smaller than the total energy of the electron in the excited state, many phonons must be generated. Note that this multiphonon emission process is temperature dependent since the electron in the excited state must overcome the barrier E_a.

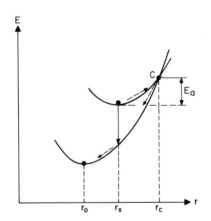

Fig. 1.4. Configuration diagram showing the radiative transition from the excited state (solid arrow) and the non-radiative path via *c* by phonon emission

1.2.2 Auger Effect

A recombining electron can transfer the energy it would have radiated to another electron in the excited state. The second electron is then raised to a higher energy. This is the Auger process. The second electron can now return to a lower energy excited state by multiple phonon emission. For example, a conduction band electron could be excited to a higher energy within the conduction band and then it would be thermalized within the band by phonon emission. In the above process, a recombination took place but no radiation has been emitted.

In the Auger effect, the energy exchange need not be confined only to carriers in similar states. The energy can be given up to carriers in other states. Examples of possible Auger exchanges are shown in Fig. 1.5.

Sometimes it is possible to identify the Auger process by its kinetics since three particles are involved: two electrons and one minority hole, or a minority

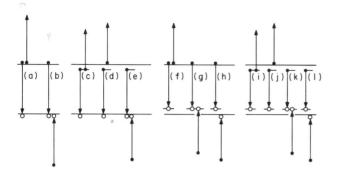

Fig. 1.5a–l. Diagram of Auger processes in a semiconductor. Processes (a), (c), (d), (f), (i), and (j) would be expected to occur in *n*-type semiconductors, while (b), (e), (g), (h), (k), and (l) would occur in *p*-type materials

electron and two holes. The probability, P_A, for an Auger event increases as either n^2p or np^2 where either the electron concentration n or the hole concentration p (or both) increase with the excitation current [1.11]. The Auger effect, in addition to decreasing the luminescence efficiency, being another pathway for recombination, also reduces the carrier lifetime $\tau_A (\tau_A = 1/P_A)$. Of all the non-radiative processes, the most tractable one experimentally is the Auger effect because, in this process, a hot carrier is generated at about twice the energy needed to excite the luminescence. The hot carrier can signal its existence by overcoming a barrier either inside the device [1.12] or at its surface [1.13]. In rare cases, the Auger effect can result in an anti-Stokes process [1.14a], one in which the hot carrier radiates its total energy, emitting photons at energies greater than the excitation intentionally supplied to individual carriers.

1.2.3 Non-Radiative Defects

It is well known that surface recombination is non-radiative, possibly because a continuum (or quasi-continuum) of states may join the conduction band to the valence band [1.14b, c]. The recombination at surface states dissipates the excess energy by phonon emission. Crystal defects such as pores, grain boundaries and dislocations may provide regions where a localized continuum of states bridges the energy gap and allows non-radiative recombination (Fig. 1.6). In the limit, a cluster of vacancies or a precipitate of impurities (e.g., a metallic phase) could also form such a non-radiative center.

Although the continuum-of-statestype of recombination center is very localized, its effect extends, on the average, over a carrier diffusion length. As carriers are depleted around such centers, they are continuously replenished by the concentration gradient, the driving force of the diffusion process.

Not all defects are recombination centers. In some instances, the local strain induced by the defect may either widen the energy gap, thus repelling

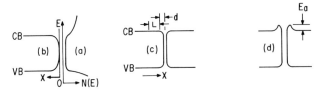

Fig. 1.6a–d. Model of non-radiative recombination via a continuum of states (a) at: (b) surface, (c) defect or inclusion (L is the electron diffusion length), (d) similar center surrounded by a barrier

both carriers, or create a local field which separates the electron-hole pairs. In either case, although the carriers are conserved, they do not recombine radiatively at such a site [1.15].

1.3 Emission Efficiency

The quantum efficiency of the luminescence process itself is called the internal quantum efficiency and is a quantity of great practical importance. It tells how well the radiative transition competes with non-radiative recombination. In an actual device, the overall efficiency (called the external quantum efficiency) is further degraded by a variety of losses (e.g., absorption and internal reflection, which affect the transmission of light from inside the device to outside).

The probability for radiative recombination P_r is related to the radiative lifetime t_r by: $P_r = 1/t_r$. The probability for non-radiative recombination P_{nr} is related to a (usually shorter) lifetime due to non-radiative recombination t_{nr} by $P_{nr} = 1/t_{nr}$. The quantum efficiency for luminescence is given by that proportion of all transitions which is radiative, i.e.:

$$\eta_q = (1/t_r)/(1/t_r + 1/t_{nr}) = 1/(1 + t_r/t_{nr}). \tag{1.2}$$

If the emitted light is partly absorbed on its way to the surface of the device, traversing a distance d, the light attenuation by absorption is $\exp(-\alpha d)$, where α is the absorption coefficient. α depends on the wavelength.

When the light reaches the surface of the device, it suffers a partial reflection due to the different refractive indices: n inside the material and ($n_{air} = 1$) outside. The reflection coefficient R is given by:

$$R = (n-1)^2/(n+1)^2. \tag{1.3}$$

The value of R for a semiconductor-air interface is typically in the range 19 to 30%. An anti-reflection coating will improve the transmission of light approaching the surface.

Since one always has $n > 1$, light approaching the surface (the air interface) at an angle θ_i with respect to the normal to the surface emerges at an angle θ_0 given by Snell's law:

$$\sin \theta_0 = n \sin \theta_i. \tag{1.4}$$

The rays which emerge at $\theta_0 = 90°$, grazing the surface, correspond to a cone of rays inside the device having a critical angle $\theta_c = \arcsin(1/n)$. All the rays approaching the surface at an angle greater than θ_c suffer total reflection and cannot escape in their first trip to the surface. The value of θ_c is typically in the range 16 to 23°.

The loss due to total internal reflectance can be minimized by resorting to a convex surface centered about a point light source. For an extended source, a roughened surface may help if the internal absorption is low, because, after reflection, the light can make several additional trips at different angles to the surface until escape is possible.

When all the above factors are taken into consideration, the external efficiency is reduced to:

$$\eta_{ext} = \eta_q(1 - R)(1 - \cos \theta_c) \exp(-\alpha d). \tag{1.5}$$

Although η_q can be close to 100% in practical LEDs, η_{ext} is usually less than a few percent.

1.4 Excitation Mechanisms

In electroluminescence, the excitation is provided by supplying either potential or kinetic energy (or both) to free carriers (mostly electrons). A forward bias at a *pn* junction or at some other barrier gives an electron the necessary potential energy. A reverse bias at the junction or a high electric field in an insulating material produces hot carriers. Sometime, the carriers must first be introduced into the insulator before they can be accelerated by the electric field. We shall also consider exotic ways of generating high fields.

1.4.1 Injection Luminescence

All the commercially available LEDs operate on the principle of forward-biased *pn* junctions. However, some exploratory work with various materials uses Schottky barriers for the injection of minority carriers.

pn Junction

A forward bias across a *pn* junction raises the potential energy of electrons in the *n*-type region, reducing the height of the barrier which prevented their spilling into the p-type region (Fig. 1.7). Similarly, the barrier which blocks the flow of holes from p-type to n-type regions is reduced by the forward bias. As soon as an overlap of electrons and holes occurs, their recombination becomes possible via processes described in previous sections. Note that the tail of the carrier's Boltzmann distribution extends to energies greater than the barrier. Therefore, some current will flow in narrow gap materials even at a low forward bias. However, since the number of free carriers increases rapidly toward the Fermi level, the current will grow exponentially with the forward bias. The current obeys the diode equation [1.16]:

$$I = I_0[\exp(qV_j/\beta kT) - 1] \tag{1.6}$$

where V_j is the voltage appearing at the junction; V_j may differ from the applied voltage V_a by the ohmic drop across the internal resistance R:

$$V_j = V_a - IR. \tag{1.7}$$

I_0 in (1.6) is the so-called "saturation current", the current obtained with a reverse bias, i.e., with V_j negative. Usually, the current does not saturate because it is dominated by the presence of generation-recombination centers [1.17] inside the pn junction and especially wherever the junction intersects a surface. In the ideal pn junction, the saturation current is controlled by the number of minority carriers which are able to diffuse to the junction; in this ideal case, I_0 depends on the density of states in the band, the energy gap, the position of the quasi-Fermi levels and the minority carrier diffusion lengths. The coefficient β is usually 1 or 2; $\beta = 1$ in the ideal diode; $\beta = 2$ when I_0 is dominated by generation-recombination centers at midgap (it can be between

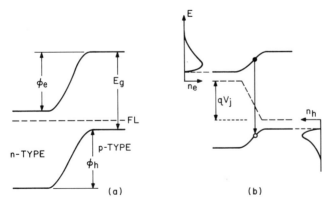

Fig. 1.7a and b. Diagram of a pn junction showing: (a) barriers φ seen by electrons and holes at equilibrium; (b) barrier reduction by a forward-bias V_j; the electron and hole distributions are labeled n_e and n_h respectively; the arrow illustrates a band-to-band recombination

1 and 2 in an intermediate case where both mechanisms are operant or when the centers are not at midgap). β appears greater than 2 when one forgets to take the internal resistance into account and uses V_a instead of V_j in (1.6).

Light-emitting diodes often have one side of the junction more heavily doped than the other. In fact, on the heavily doped side the material is usually degenerate, i.e., the Fermi level is inside the band by an amount ξ. As can be seen in the illustration of Fig. 1.8, the carriers in the degenerately doped band have the advantage of a reduction of barrier height by ξ compared to those in the more lightly doped side. A second benefit of heavy doping in the n^+ region of Fig. 1.8 is that band-filling raises the threshold energy for absorption by ξ, so that the emitted light can traverse the n^+ region without much absorption. When both sides are degenerately doped, an additional phenomenon must be considered: the relative shrinkage of the energy gap on either side due to band perturbation effects [1.18].

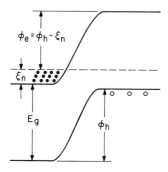

Fig. 1.8. Diagram of a *pn* junction between a degenerately doped *n*-type and a lightly doped *p*-type semiconductor

Heterojunction

In order to control which carrier is injected into the luminescent material, one can resort to a heterojunction (Fig. 1.9). The source of injected carriers is a material having a wider gap than that of the luminescent region. Then, the

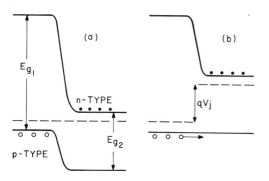

Fig. 1.9a and b. Diagram of a hetero-junction (a) at equilibrium, (b) with a forward bias V_j

asymmetry of the barriers seen by electrons and holes assures a high injection efficiency from the wider gap to the narrower gap layer [1.19, 20]. Although heterojunctions can be made between very different materials, the most successful devices are those made from different compositions of miscible alloys having similar lattice constants at fabrication temperature – a precaution which minimizes dislocations and interfacial stress – e.g., $Al_{1-x}Ga_xAs$ with two different values of x on either side of the junction [1.21].

Schottky Barrier

A Schottky barrier usually occurs at the surface of a semiconductor in contact with a metal. Such a barrier can act as a *pn* junction: whenever surface states induce an inversion layer, the surface has the opposite conductivity type than the bulk. In effect, this is equivalent to a *pn* junction immediately below the surface. Figure 1.10 illustrates two cases of inversion layer in *n*- and *p*-type materials, respectively. A forward bias tends to flatten the bands, allowing the injection of minority carriers into the bulk. The minority carriers can then recombine radiatively just as they would in a *pn* junction.

One concern with a forward-biased Schottky barrier is the relative ease with which majority carriers (electrons in Fig. 1.10a) can flow to the metal electrode, thus reducing the injection efficiency and, therefore, the luminescence efficiency. The injection efficiency depends on the relative population of majority carriers at the surface and in the bulk. A strong inversion on a lightly doped crystal is the most favorable combination for high injection efficiency. This can be readily visualized by examining Fig. 1.10. For example, if the *n*-type layer in Fig. 1.10a is lightly doped, the Fermi level is farther from the conduction band in the bulk than it is from the valence band at the surface. In other words, the barrier to electrons will be higher than the barrier to holes. There will be fewer electrons flowing from the semiconductor to the metal than holes flowing in the opposite direction.

Schottky barriers are easy to make: a point contact or an evaporated layer on the surface forms a Schottky barrier. A thin metal layer can be simultaneously conducting and partly transparent (with a flat spectral transmission). Although such a structure facilitates the emission of light generated near the surface, it is not as practical as a *pn* junction or a heterojunction because of the limited transmission of the metal electrode and because of the relatively

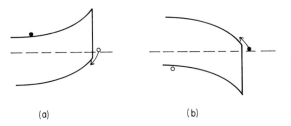

(a)　(b)

Fig. 1.10a and b. Diagram of Schottky barriers to *n*-type (a) and *p*-type (b) material. The arrows show minority carrier injection during forward bias

low minority carrier injection efficiency. On the other hand, it is a convenient technique for a fast experiment with materials in which a good control of *pn* junction technology has not been achieved.

MIS Structure

When surface charges are insufficient to induce the desired inversion layer, one can induce surface charges by a capacitive coupling to the surface through an insulator. In a metal-insulator-semiconductor (MIS) structure, the surface charge on the semiconductor is determined by the potential applied to the metal [1.22]. For the present discussion, we shall call ZnS and other luminescent insulators "semiconductors" to distinguish them from the wider gap insulator placed on their surface. An immediate benefit of the MIS structure is that the band bending can be controlled by the applied voltage. Thus, an inversion layer can be induced by one polarity, then the carriers accumulated at the surface can be injected by reversing the polarity of the metal electrode (Fig. 1.11). This method has been used to obtain electroluminescence in GaAs without a *pn* junction [1.23]. Another example of this mode of excitation is the UV electroluminescence of GaN [1.24].

If the insulator is made very thin (less than about 100 Å), electrons can tunnel across the insulator [1.25].

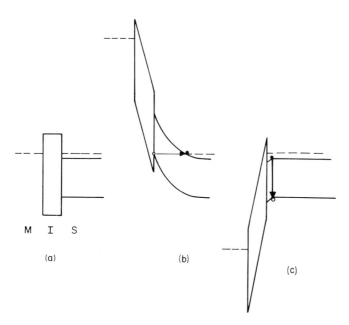

Fig. 1.11a–c. Band structure at MIS transition: (a) without bias, (b) metal negatively biased for hole generation, (c) metal positively biased for recombination

If the insulator is so thin that the exponentially decaying probability of finding the electron extends from both sides of the insulator without too much attenuation, the electron can readily reappear on the other side of the insulator without loss of energy. This tunneling property of thin layers can be used to inject electrons into the conduction band of a p-type semiconductor (Fig. 1.12). However, with rare exceptions [1.26], it is not suitable for hole injection (electron extraction by tunneling from the valence band) because, with n-type material, there is a high probability for electrons to simultaneously tunnel out of the conduction band.

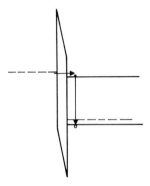

Fig. 1.12. Electron injection by tunneling through a thin insulator

1.4.2 Radiative Tunneling

The process of radiative tunneling, or photon-assisted tunneling, or tunneling-assisted photon emission or diagonal tunneling was first identified in a GaAs pn junction [1.27–29]. An electron from the conduction band tunnels into the gap where it completes radiatively the transition to the valence band (or to an empty state in the gap). The wave functions of both electrons and holes have exponentially decaying components beyond the sloping band edges, which act as one wall of a potential well (Fig. 1.13). An overlap of these exponential probability distributions makes the radiative recombination in the space charge region possible; the probability of such a recombination increases with the overlap. This process is possible only when the semiconductor is degenerately doped on both sides of the pn junction so that the width of the depletion region is less than about 100 Å when a forward bias is applied.

The radiative tunneling process can be easily identified. Most of the tunneling occurs at the quasi-Fermi levels. Hence, the peak of the emission spectrum shifts with the applied voltage. Furthermore, the emission spectrum peaks at photon energies lower than those obtainable in the same material by other excitation methods such as photoluminescence or cathodoluminescence. In the case of a band-to-band transition, the emission peaks at $h\nu = qV_j$; if the transition is from the conduction band to an acceptor at energy E_a above the valence band, then $h\nu = qV_j - E_a$ [1.28].

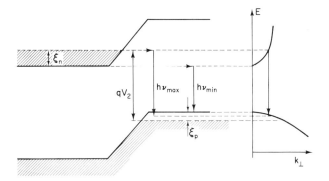

Fig. 1.13. Radiative tunneling across a *pn* junction with forward bias V_2

1.4.3 Breakdown Luminescence

In the presence of an electric field, an electron can gain enough kinetic energy to impact excite other electrons either from the valence band or from impurities. In this way, the number of free electrons can be multiplied by successive collisions. If both electrons and holes participate in the breakdown process (and both are multiplied), the process is called "avalanche" breakdown [1.30]. Unlike the multiplication process which builds up a positive space charge where impact ionization has occurred, in the avalanche mechanism, the material remains neutral and tends to sustain a uniformly distributed high field. Since the electrons and holes flow in opposite directions, the holes provide a regenerative feedback (producing more free electrons near the cathode). Avalanching builds up the carrier concentration until some rate-limiting process saturates the current (e.g., ohmic drop at a contact), or until the material vaporizes.

The kinetic energy needed for pair creation by impact excitation across the gap depends on the energy gap E_g and the effective masses of the carriers. Both energy and momentum must be conserved. The threshold energy is given by

$$E_t = \frac{m_h^* + 2m_e^*}{m_h^* + m_e^*} E_g \qquad (1.8)$$

where m_h^* and m_e^* are the effective masses of holes and electron, repectively. If the two masses are assumed to be equal, the threshold energy reduces to:

$$E_t = (3/2)E_g. \qquad (1.9)$$

Impact excitation from a deep center would require less kinetic energy because of the extensive spread of the initial state in momentum space.

In a uniformly doped material, the threshold energy can be achieved either by adjusting the applied voltage V_a, or by selecting the distance d between the electrodes such that:

$$V_a^2/d^2 = 2E_t/m\mu^2 \qquad (1.10)$$

where μ is the mobility. Of course, qV_a must, itself, be at least equal to E_t. If the material is too thin, the electrons will tunnel through it without useful impact excitation.

The pairs created in a breakdown process can recombine radiatively. Most of the carriers are rapidly thermalized to the band edges. Hence, one might expect the emission spectrum to peak at the bandgap energy. Because of the high electric field, some carriers can make a radiative tunneling transition. which broadens the spectrum to lower photon energies. Hot carriers can recombine also, generating photons of energy greater than bandgap. An accurate description of the avalanche-breakdown luminescence spectrum fits the following expression [1.31]:

$$U(v) = A[1 - \mathrm{erf}\,(hv/C)] \qquad (1.11)$$

where U is the light intensity, A is a constant and C depends on the ratio of carrier velocity to sound velocity, on the kinetic energy of the hot carrier, and on the details of collisional losses by the hot carriers.

The higher energy edge of the emission spectrum can be readily reabsorbed by the semiconductor. Therefore, it is observed only in specially prepared structures in which the breakdown occurs at or very near the surface. Avalanche emission has been studied in Si [1.32], Ge [(1.33], SiC [1.34], and $GaAs_{1-x}P_x$ [1.35].

Some of the early commercial light emitting diodes were Si avalanche diodes emitting white light. Although their efficiency was low, they were very fast and could be used to test the response time of photomultipliers.

In semiconductors, the most convenient structure for observing the breakdown luminescence is either a shallow pn junction or a Schottky barrier to which a reverse bias is applied. Since the depletion layer is usually thin ($< 1\,\mu m$), the high field needed for breakdown ($\sim 2 \times 10^5 V/cm$) can be obtained at a low voltage.

In insulators such as electroluminescent phosphors, on the other hand, breakdown occurs at higher fields ($\sim 10^6 V/cm$). However, insulators have no free carriers to accelerate. Therefore, the high field must first generate the free carriers. These are produced either by Zener tunneling [1.25] (electron tunneling from the valence band to the conduction band) or, more probably by excitation from a trap (Frenkel-Poole effect) or over the Schottky barrier at the cathode. Note that Zener tunneling is not strongly temperature dependent —only the temperature dependence of the gap is involved—whereas both Frenkel-Poole and Schottky emissions are strongly temperature dependent (see Table 1.1). This is due to the fact that thermally heating the carriers helps them overcome the Schottky barrier at the electrode interface or the barrier

Table 1.1. I(V) Characterization of various transport processes

Conduction process in semiconductors and insulators	Dependences of current on voltage and temperature
Injection	$\exp(qV/kT)$
Tunneling	$V \exp(-b/V^{\frac{1}{2}})$ or $V^2 \exp(-b/V)$
Schottky emission	$T^2 \exp[(-B+aV^{\frac{1}{2}})/kT]$
Frenkel-Poole	$V \exp[(-B+2aV^{\frac{1}{2}})/kT]$
Space-charge-limited current	V^2/d^3

a, b, B, q, k are constants; d is the distance between electrodes.

located at each potential well. This is thermionic emission (which does not occur in Zener tunneling).

In EL phosphors, two excitation mechanisms are possible but they are difficult to distinguish: 1) raising a neutral center, such as a rare earth atom, to an excited state by direct impact of the center; or 2) pair generation with immediate trapping of an electron-hole pair by the luminescent center. Since impact excitation and carrier multiplication occur after some mean free path – not necessarily the same for the two processes – the carrier concentration and the luminescence intensity should increase exponentially with the thickness of the phosphor [1.36].

If electrons are injected into a region of an insulator where no compensating charge is present, a space charge will build up near the cathode. The potential distribution will become nonlinear and the field will maximize at the anode. The current will be limited by the accumulated space charge. It can be shown that the current is given by [1.37]:

$$I = \varepsilon \mu V^2 / d^3 \tag{1.12}$$

where ε is the dielectric constant, μ is the electron mobility and d is the distance between cathode and anode. The $I(V)$ characteristics due to the various processes discussed above are summarized in Table 1.1.

The impact ionization of neutral impurities may also produce a trapped space charge which will result in a non-uniform field distribution. Trap filling (and emptying) at some critical voltage may cause the current to rise rapidly over a narrow range of voltage beyond which the square law is obeyed [1.37].

1.4.4 Luminescence from Traveling High Field Domains

Several phenomena convert a relatively low applied electric field into domains of intense electric field. These domains travel along the specimen (usually from cathode to anode) and generate a high density of electron-hole pairs either by impact excitation or by Zener tunneling. In some structures, the domain interacts with *pn* junctions to cause either injection or breakdown. Although these effects have been only laboratory curiosities until now, it is too early to discount their practical potential.

Luminescence Induced by the Gunn Effect

One method of creating impact ionization in GaAs is to generate Gunn domains in lightly doped *n*-type material [1.38, 39]. Gunn domains are formed as follows. Electrons are accelerated by the electric field ($> 2.2 \times 10^3$ V/cm). When the hot electrons have sufficient kinetic energy to populate the indirect valley where their efffective mass is m_2^* (see Fig. 1.14(a)), their mobility suddenly decreases to $\mu_2 = q\tau/m_2^*$ (τ is the carrier relaxation time) and a higher resistivity region or "domain" results (Fig. 1.14(b)). Most of the applied voltage develops across the high-resistivity domain. The domain being thin, the electric field in

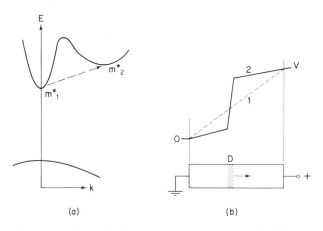

(a)   (b)

Fig. 1.14a and b. Excitation by Gunn domain: (a) transfer of hot electron to indirect low mobility valley; (b) potential distribution across specimen: 1 — before and 2 — after formation of traveling domain *D*

this region exceeds the impact-ionization value (2×10^5 V/cm) and generates many electron-hole pairs. Their radiative recombination forms a light source which propagates with the Gunn domain [1.38]. When the field is turned off (or in the wake of the domain), the hot carriers thermalize to the edges of the bands and produce a more intense burst of direct radiative transition. When a Fabry-Perot cavity is provided, coherent radiation is obtained [1.40].

Visible emission from Gunn domains could be expected from wider gap materials having higher valleys where the mobility is reduced. Such materials may be $GaAs_{1-x}P_x$ or $Ga_{1-x}Al_xAs$ where x approaches the crossover composition.

Luminescence Induced by the Acoustoelectric Effect

Another means of generating a traveling high-field domain is the acoustoelectric effect [1.41–43].

The acoustoelectric effect is a cooperative phenomenon between electrons and phonons. Under the influence of an electric field, when the carriers are accelerated to a velocity comparable to the velocity of sound in the material, they transfer some of their kinetic energy to the lattice in the form of phonons. As the electron tends to exceed the sound velocity, it gives up more energy to phonons. Hence, after an initial acceleration, the velocity of the electron, on the average, saturates at the sound velocity v_s. The process of energy transfer continues along the statistical path of the electron, thus building up the intensity of the accompanying phonons. In a uniform n-type semiconductor, the phonon buildup which starts at the cathode is the one which can grow the most, since it has the longest path to traverse.

A velocity saturation corresponds to a decrease in mobility, and, therefore, to an increase in resistivity. With a constant voltage applied across the specimen, the field distribution along the crystal evolves from an initially uniform field to a stepped electric field: as electrons near the cathode reach their saturation velocity, the local resistivity increases, forcing the local field to grow at the expense of the field in adjacent regions (Fig. 1.15). The high-field region, called the "acoustoelectric domain", travels with the velocity of sound.

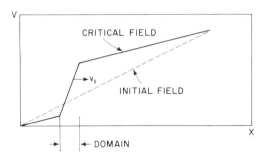

Fig. 1.15. Potential distribution along the semiconductor before and after acoustoelectric domain formation

Note that as soon as the domain is formed, an increasing fraction of the applied voltage is developed across the domain, where more electrical energy is transformed into phonons. Hence, the domain tends to grow until it eventually saturates when a large portion of the applied voltage appears across the domain, while the field outside the domain drops to a critical value which is that field at which the electrons travel at the sound velocity.

Since the initial field is in excess of the critical field before the domain is formed, the carriers can travel at a velocity greater than v_s. The current, which still obeys Ohm's Law, is therefore high. However, while the domain travels along the semiconductor, the carriers slow down to the saturated velocity causing the current in the external circuit to drop below the initial value. The corresponding $I(V)$ characteristic is shown in Fig. 1.16 and time dependences of current and voltage are shown in Fig. 1.17.

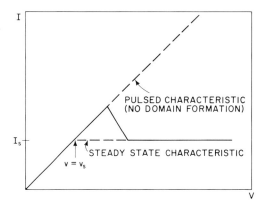

Fig. 1.16. $I(V)$ characteristics of semiconductor exhibiting the acoustoelectric effect. Under pulsed excitation for a time too short to allow domain formation, the ohmic characteristic is obtained. For a long excitation, the steady-state characteristic is obtained, the current saturating at $I_s = qnv_s$ (n is the carrier concentration). For an intermediate duration of excitation the solid curve is obtained

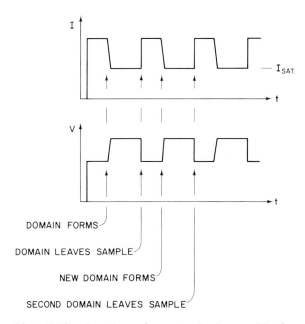

Fig. 1.17. Time dependence of current and voltage resulting from the acoustoelectric effect

When the domain reaches the end of the specimen, the field inside the crystal suddenly increases and the current surges back to its ohmic value. As soon as the domain has been swept out, a new domain forms at the cathode and the current again drops to the saturation value. With a long enough excitation ($t > l/v_s$; l = length of specimen), a number of such cycles can be obtained (Fig. 1.17). If the specimen is not driven at constant voltage (due to a finite circuit resistance), the voltage across the specimen increases when the current decreases; hence, the voltage wave shape is a mirror image of the current wave shape.

It must be pointed out that, although energy exchange between electrons and phonons occurs in all semiconductors, the acoustoelectric domain can be formed only in piezoelectric semiconductors. Therefore, the occurrence of acoustoelectric domains depends on crystallographic orientation, just as does the piezoelectric effect (e.g., it is maximum along the [110] axis in GaAs and $GaAs_{1-x}P_x$).

A burst of light emission has been observed in the dumbbell-shaped specimen of Fig. 1.18 when the acoustoelectric domain reaches the end of the bar [1.44]. This emission, the spectrum of which corresponds to near-gap recombination, can be explained as follows: the electric field in the domain separates electrons and holes into a moving dipole (here, the holes are field-ionized minority carriers) [1.45–47]; when the electric field collapses at the flared end of the specimen near the anode, the electron and hole clouds are allowed to overlap and recombine radiatively.

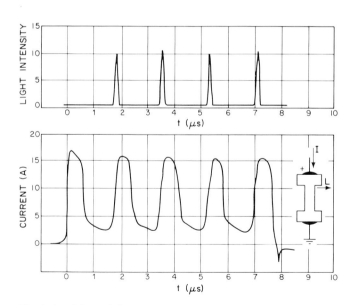

Fig. 1.18. Light emission in n-type GaAs when the acoustoelectric domain reaches the end of the bar in the dumbbell-shaped specimen [1.42]

If an array of electrically floating *pn* junctions is built on the semiconductor bar, each *pn* junction can be caused to emit in turn as the domain passes by [1.48–50].

A floating *pn* junction [1.51] is a structure (shown in Fig. 1.19) where the net current through the *pn* junction is zero. In the presence of a potential distribution such that the equipotential planes intersect the *pn* junction, part of the junction is reverse biased and the rest of the junction is forward biased. The ratio of forward-biased-to-reverse-biased areas adjusts itself in such a manner that the forward current is equal to the reverse current. Since the average reverse-current density is much lower than the average forward-current density, the reverse-biased area is much larger than the forward-biased area. The forward-biased portion of a *pn* junction can emit by injection luminescence, while the reverse-biased portion can generate light in the breakdown mode.

A floating junction is equivalent to an "*npn*" hook (Fig. 1.19(b)); one side is reverse biased and the other is forward biased. To maintain charge neutrality in the *p*-region, the electrons injected into the *p*-region by the forward-biased junction are compensated by holes collected at the reverse-biased junction (these holes may result from electrons tunneling to the conduction band in the high-field region). The radiative recombination is very efficient in the case of two-carrier flow to the *p*-type region of GaAs and $GaAs_{1-x}P_x$ devices [1.49]. In the case of breakdown luminescence, which has been observed with floating Cu_2S-CdS heterojunctions [1.46] (Fig. 1.19(c)), the two carriers are formed in the high-field region, where their encounter time is very short. The holes flow to the *p*-type region, where they are majority carriers. In the forward-biased

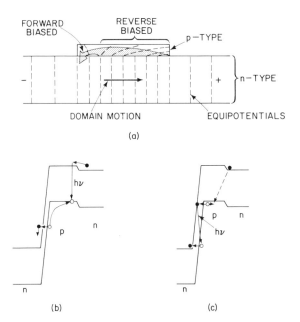

Fig. 1.19a–c. Current flow in a floating *p-n* junction

Cu₂S-CdS junction, charge neutrality must be obtained by non-radiative recombination (or tunneling), since injection luminescence was not found in this Cu₂S-CdS system.

It is conceivable that the principle of the acoustoelectric domain inducing a spatially sweeping injection luminescence along an array of *pn* junctions may someday result in a practical solid-state flying-spot scanner.

1.5 Psychophysical Considerations

The variety of luminescent materials and phenomena is so great that sometimes the user is puzzled by the available choice. If the light is to activate an electronic system, the choice of the light source should be guided by the optimal match between the characteristics of the light source and those of the detector. However, if the emitted light is to interact with the human eye (for example in a display panel) one must be aware of the psychophysical characteristics of the eye as a detector. Hence, in the following few paragraphs, we shall review the response of the average human eye.

Color vision is provided by the cones in the retina whereas sensitivity to low light levels is provided by the rods. At low light levels, there is a loss of color perception.

The threshold of response to visual stimuli depends on the background illumination. The minimum detectable stimulus corresponds to about 60 quanta of blue-green light at 510 nm impinging on the cornea [1.52].

The rods in the retina provide the "scotopic response". It occurs after the eye has been dark adapted to at most 3×10^{-5} nits (Cd/m²). Complete dark adaption requires about 45 min. The spectrum of luminous efficiency as a function of wavelength is shown in Fig. 1.20.

The "photopic" response is due to the cones. It occurs after the eye has been adapted to a background illumination of at least 3 nits, i.e., in the light-adapted state. This adaption requires about two minutes when the luminescence is increased. Fig. 1.20 shows the spectral response of photopic vision.

Between the light-adapted (photopic) and dark-adapted (scotopic) states, the response is continuously variable; the spectral response shifts towards the blue when the illumination decreases. This condition is known as the mesopic state, which occurs in the range 3nt to 3×10^{-5}nt. For most practical applications it is the photopic state which is involved.

Color perception is characterized by three subjective attributes:
1) luminance – often called brightness;
2) hue – the distinguishable color (pure colors are directly related to wavelength);
3) saturation or chroma – which measures how intense the color is (or how undiluted it is by white light which would render this color pale).
Any color sensation can be reproduced by the judicious combination of three monochromatic components (for example, red, green and blue lights). The

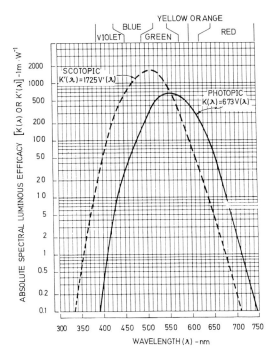

Fig. 1.20. Absolute spectral luminous efficacy of the human eye as a function of wavelength

Commission Internationale de l'Eclairage (CIE) has adopted a standard of colorimetry which represents the attributes of color by a three-dimensional diagram. The Cartesian vectors of this tridimensional diagram are derived from three idealized primaries which are non-monochromatic as shown in Fig. 1.21. The coordinates to be used to represent a color are calculated by a procedure to be described later on. The values of x,y and z are calculated in such a way that their sum is unity $(x+y+z=1)$. This property makes a three-dimensional plot unnecessary. It is sufficient to identify x and y to automatically imply z. In other words, the color map may be expressed as a two-dimensional projection into the xy plane. This is the standard chromaticity diagram shown in Fig. 1.22.

The upper arc is the locus of saturated colors. All the colors and shades that the eye can resolve are enclosed between the area of saturated colors and the straight line labeled "Magenta". The central region appears white. A black body assumes at different temperatures various shades of whiteness as shown by the arc labeled "black body locus".

Now returning to the tristimulus values of Fig. 1.21, the curve labeled $\bar{y}$ is the apparent intensity of a constant power of a monochromatic light shown on the abscissa; curve $\bar{y}$ is normalized at a peak at 550 nm. The reader will readily recognize that curve $\bar{y}$ is the photopic response of the eye. If a color is generated by a monochromatic source, one calculates its chromatic coordinates

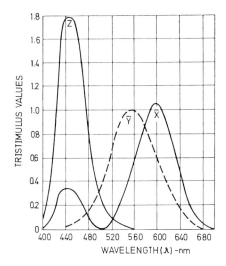

Fig. 1.21. CIE standard color mixture curves

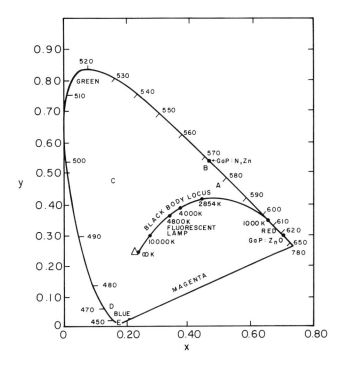

Fig. 1.22. CIE chromaticity diagram. The points ABCDE were obtained with GaN LEDs, the spectra of which are reported in Fig. 9 of [1.56]. The point Δ corresponds to the spectrum in Fig. 1.24

by the ratios:

$$x = X/(X + Y + Z)$$
$$y = Y/(X + Y + Z)$$
$$z = Z/(X + Y + Z)$$

where the tristimulus values X, Y, and Z are read at the chosen wavelength on Fig. 1.21. The ratios are the means of normalizing x, y, and z so that their sum be unity.

If the color is generated by non-monochromatic sources, the three tristimulus values are calculated by the following integrations over the entire visible spectrum:

$$X = \int \phi_\lambda x(\lambda) d\lambda$$
$$Y = \int \phi_\lambda y(\lambda) d\lambda$$
$$Z = \int \phi_\lambda z(\lambda) d\lambda$$

where ϕ_λ is the total spectral radiant flux. The chroma coordinates are determined by the previously defined ratios.

Color discrimination by the eye is much better in the lower portion of the chromaticity diagram than in the upper portion, for example, the sensation of a given shade of green can be obtained from a larger area of the chromaticity diagram than a given shade of blue or of pink. In order to equalize this error in color discrimination, various new diagrams have been devised which stretch the x axis and compress the upper part of the y axis.

Loebner [1.53] describes the many forms of color perception abnormalities which affect about 5% of the male population and about 0.4% of the female population. Monochrome displays most suitable to color abnormals should be either yellow or white.

The eye is capable of adapting to a 10^7 range in light level. Most of the adaptation takes place within the retina. Changes in the size of the pupil account for only a 20:1 range in light level adaptation. The purpose of increasing the f number of the eye is primarily to increase the sharpness of the image.

The chromatic aberration of the eye (Fig. 1.23) causes the shorter wavelength of light to be focused much closer to the lens than the longer wavelengths [1.54]. In fact, blue light is never brought to a sharp focus; consequently, the resolving power in the blue is poor. Another consequence of the axial chromatic aberration is the difficulty in focusing on two objects in the same place when their colors are so different that an accomodation of more than 0.4 diopter is required. Thus, if the ambient illumination is mostly yellow-green (~ 570 nm), it may be difficult to focus on either the red ($\lambda > 650$ nm) or the blue ($\lambda < 500$ nm) portions of a picture. Chromatic aberration may account for the eye fatigue experienced by viewers of red LED displays who wear bifocal or trifocal correcting lenses (i.e., those viewers who already have difficulty with focus accommodation). It should be pointed out that a special lens system has been designed to completely compensate the chromatic aberration of the human eye [1.54].

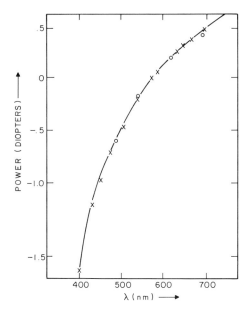

Fig. 1.23. Axial chromatic aberration of the human eye measured in diopters and normalized (0 diopter) at 578 nm. The crosses are the data of *Bedford* and *Wyszecki* [1.54], the circles are data of *Wald* and *Griffin* [1.57]

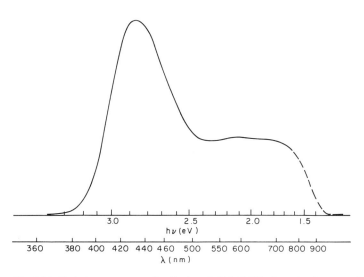

Fig. 1.24. Emission spectrum of a Zn-doped GaN LED. Its color is mapped as point Δ on the chromaticity diagram (Fig. 1.22). The dashed portion in the infrared is dominated by the spectral cutoff of the photomultiplier (RCA 31025)

In cases where contrast rather than resolution is important, e.g., visibility of a large display, another factor must be considered: The spectral responses of red and green sensitive cones have a considerable overlap, whereas the spectral overlap with the blue-sensitive cones is small. Hence, in artificial ambient illumination, which is already enriched with red light, the red LED display has a much lower contrast than a blue LED display of equal luminance.

If one assumes comparable efficiency and comparable manufacturing economics for EL devices, the most useful material would be the one capable of emitting white light. Its color output could be tailored by the choice of impurities within a broad spectral range or by filters. All the wide gap materials ($Eg > 3.0$ eV) are potential candidates. The most encouraging results have been obtained with $ZnS_{1-x}Se_x$ emitting in the yellow to blue range [1.55] and with GaN, the emission range of which extends from reddish to UV [1.56]. GaN has produced not only the nearly saturated colors shown along the periphery of the chromaticity diagram (Fig. 1.22) but also white light — for example, the spectrum of Fig. 1.24 which is mapped as point Δ in the chromaticity diagram. The incomplete color saturation of GaN is due to the broad half-width of the emission ($\Delta\lambda > 650$ Å). Better saturation but over a narrower chromatic range is obtained with GaP and $GaAs_{1-x}P_x$. *Loebner* [1.53] has speculated on the possibility of using the direct gap alloy $In_{1-x}Ga_xN$ to make EL devices capable of emitting at virtually any coordinate inside the chromaticity diagram.

I am grateful to *H. Kressel*, *C. J. Nuese* and *D. Redfield* for a critical reading of the manuscript.

References

1.1   D. G. Thomas, J. J. Hopfield, C. J. Frosch: Phys. Rev. Lett. **15**, 857 (1965)
1.2   J. J. Hopfield, D. G. Thomas, M. Gershenzon: Phys. Rev. Lett. **10**, 162 (1963)
1.3   F. Williams: J. Luminesc. **7**, 35 (1976)
1.4   J. S. Prener, F. E. Williams: Phys. Rev. **101**, 1427 (1956)
1.5   F. E. Williams: J. Phys. Chem. Solids **12**, 265 (1960)
1.6   H. Reiss, C. S. Fuller, F. J. Morin: Bell Syst. Tech. J. **35**, 535 (1956)
1.7   J. J. Hopfield: *II-VI Semiconducting Compounds*, ed. by D. G. Thomas (Benjamin, New York, Amsterdam 1967) p. 796
1.8   T. N. Morgan, H. Maier: Phys. Rev. Lett. **27**, 1200 (1971)
1.9   T. N. Morgan, B. Welber, R. N. Bhargava: Phys. Rev. **166**, 751 (1968)
1.10 C. H. Henry, P. J. Dean, J. D. Cuthbert: Phys. Rev. **166**, 754 (1968)
1.11 J. S. Blakemore: *Semiconductor Statistics*, (Pergamon, Oxford, Elmsford, NY 1962) p. 214
1.12 M. I. Nathan, T. N. Morgan, G. Burns, A. E. Michels: Phys. Rev. **146**, 570 (1966)
1.13 J. I. Pankove, L. Tomasetta, B. F. Williams: Phys. Rev. Lett. **27**, 29 (1971)
1.14a R. Conradt, W. Waidelich: Phys. Rev. Lett. **20**, 8 (1968)
1.14b A. Many, Y. Goldstein, N. B. Grover: *Semiconductor Surfaces*, North Holland (1965), p. 434, Amsterdam
1.14c S. G. Davison, J. D. Levine: Solid State Physics–Advances in Research and Applications **25**, 1 (1970)
1.15 W. Heinke, H. J. Queisser: Phys. Rev. Lett. **33**, 1082 (1974)
1.16 W. Shockley: Bell Syst. Tech. J. **28**, 435 (1949)

1.17 C. T. Sah, R. N. Noyce, W. Shockley: Proc. IRE **45**, 1228 (1957)
1.18 J. I. Pankove: *Progress in Semiconductors*, Vol. 9, ed. by A. F. Gibson, R. E. Burgess (Heywood & Co., London 1965) p. 48
1.19 H. Kroemer: Proc. IRE **45**, 1535 (1957)
1.20 A. G. Fisher: *Luminescence of Inorganic Solids*, ed. by P. Goldberg (Academic Press, New York 1966) p. 541
1.21 H. Kressel: J. Electron. Mat. **4**, 1081 (1975)
1.22 A. S. Grove: *Physics and Technology of Semiconductor Devices* (Wiley & Sons, New York, London, Sydney 1967) Part III;
       S. M. Sze: *Physics of Semiconductor Devices* (Wiley & Sons, New York, London, Sydney, Toronto 1969) pp. 425–504
1.23 C. N. Berglund: Appl. Phys. Lett. **9**, 441 (1966)
1.24 J. I. Pankove, P. E. Norris: RCA Review **33**, 377 (1972)
1.25 I. Giaever: Phys. Rev. Lett. **5**, 147 (1960)
1.26 M. D. Clark, S. Baidyaroy, F. Ryan, J. M. Ballantyne: Appl. Phys. Lett. **28**, 36 (1976)
1.27 J. I. Pankove: Phys. Rev. Lett. **9**, 283 (1962)
1.28 J. I. Pankove: J. Appl. Phys. **35**, 1890 (1964)
1.29 H. C. Casey, D. J. Silversmith: J. Appl. Phys. **40**, 241 (1969)
1.30 K. G. McKay: Phys. Rev. **94**, 877 (1954)
1.31 J. Schewchun, L. Y. Wei: Solid State Electron. **8**, 485 (1965)
1.32 A. G. Chynoweth, K. G. McKay: Phys. Rev. **102**, 369 (1956)
1.33 A. G. Chynoweth, H. K. Gummel: J. Phys. Chem. Solids **16**, 191 (1960)
1.34 G. F. Kholuyanov: Sov. Phys. Solid State **3**, 2405 (1962)
1.35 M. Pilkhun, H. Rupprecht: J. Appl. Phys. **36**, 684 (1965)
1.36 J. J. Hanak: Proc. 6th Intern. Vacuum Congress, 1974; Japan. J. Appl. Phys. Suppl. 2, Pt. 1 (1974)
1.37 M. A. Lampert, P. Mark: *Current Injection in Solids* (Academic Press, New York, London 1970)
1.38 J. S. Heeks: IEEE Trans. ED-**13**, 68 (1966)
1.39 P. D. Southgate: J. Appl. Phys. **38**, 6589 (1967)
1.40 P. D. Southgate: IEEE J. QE-**4**, 179 (1968)
1.41 P. O. Sliva, R. Bray: Phys. Rev. Lett. **14**, 372 (1965)
1.42 C. Hervouet, J. Lebailly, P. Leroux-Hugon, R. Veillex: Solid State Commun. **3**, 413 (1965)
1.43 H. Hayakawa, M. Kikuchi, Y. Abe: Japan. J. Appl. Phys. **5**, 734, 735 (1966)
1.44 A. Bonnot: Compt. Rend. **263**, 388 (1966)
1.45 S. S. Yee: Appl. Phys. Lett. **9**, 10 (1966)
1.46 N. I. Meyer, N. H. Jørgensen, I. Balslev: Solid State Commun. **3**, 393 (1965)
1.47 A. Bonnot: Phys. Stat. Sol. **21**, 525 (1967)
1.48 B. W. Hakki: Appl. Phys. Lett. **11**, 153 (1967)
1.49 J. I. Pankove, A. R. Moore: RCA Review **30**, 53 (1969)
1.50 Y. Nannichi, I. Sakuma: J. Appl. Phys. **40**, 3063 (1969)
1.51 A. R. Moore, W. M. Webster: Proc. IRE **43**, 427 (1955)
1.52 C. H. Graham, ed.: *Vision and Visual Perception* (Wiley & Sons, New York 1965) p. 154
1.53 E. E. Loebner: Proc. IEEE **61**, 837 (1973)
1.54 R. E. Bedford, G. Wyszecki: J. Opt. Soc. Am. **47**, 564 (1957)
1.55 R. J. Robinson, Z. K. Kun: Private communication
1.56 J. I. Pankove: J. Luminesc. **7**, 114 (1973)
1.57 G. Wald, D. R. Griffin: J. Opt. Soc. Am. **37**, 321 (1947)

2. Group IV Materials (Mainly SiC)

Y. M. Tairov and Y. A. Vodakov

With 25 Figures

2.1 Background

All semiconducting substances of group IV in the Mendeleyev periodic system, i.e., diamond, silicon, germanium, silicon carbide (SiC), are indirect gap materials. Therefore, radiative interband transitions due to the recombination of a free electron and a free hole can occur in these substances only with the emission of phonons (or their absorption) and are less likely than in the direct-band semiconductors. The radiative recombination of a free carrier and a bound carrier of the opposite polarity on the impurity center is more likely and radiative no-phonon transitions may appear. When the ionization energy of an impurity atom is increased, the efficiency of such luminescence (L) and its temperature stability may increase as well. It is natural that in such narrow-band semiconductors as Ge ($E_g = 0.803$ eV) and Si ($E_g = 1.21$ eV) the effective L is hardly to be expected at temperatures close to room temperature [2.1–2.4], and these materials are unlikely to be of interest with respect to electroluminescence (EL). Therefore, although reverse-biased silicon diodes were among the first intentionally produced EL diodes, they did not find any significant practical application.

On the contrary, in such broad-band semiconductors of this group as diamond ($E_g = 5.49$ eV) and SiC ($E_g \sim 3$ eV) an efficient L is observed even at temperatures greatly exceeding room temperature. Depending on the impurity composition, the spectral region of L may cover the whole visible and near infrared regions, while in diamond it may extend into the ultraviolet up to 230 nm. The most detailed studies by *Dean* [2.5, 6] of the L of natural and synthetic diamond revealed that the L is due to the annihilation of bound excitons, the recombination of donor-acceptor pairs (DAP) [2.5] and the annihilation of free indirect excitons [2.6, 7]. The more recent studies of diamond have made these results more accurate [2.8, 9]. Since it is very hard to synthesize large semiconducting diamonds (and they are very rare in nature) the generation of conducting layers on natural insulating diamonds has been attempted by ion implantation [2.10]. The ionization energy of nitrogen, the most common donor impurity in diamond, is about 4 eV. This value is too high to obtain conducting n-type layers when doping diamond with nitrogen. Therefore in the process of ion implantation in diamond, particular attention should be paid to injecting other donors, e.g., P, Sb, Li. However, in these ion-implanted layers of diamond one can succeed sometimes in observing an efficient

EL in the blue region of the spectrum possibly due to the formation of *pn* junctions. The nature of EL in diamond bulk crystals which was studied in [2.11, 12] is not yet well understood. At any rate, good injecting *pn* junctions have not been realised reproducibly in diamonds thus far. In SiC, on the contrary, both electron and hole conduction are easily realized and *pn* junctions, even spontaneously produced, may generate very intense EL. It is in such unintentionally produced *pn* junctions in SiC which appeared sometimes in crystals grown for abrasive use, that *Losev* discovered the phenomenon of injection EL in 1923 [2.13]. It is only in the last 10 to 15 years, when one learned to create *pn* junctions in SiC [2.14–18], that the development of EL devices based on this principle began.

2.2 Physical Properties of SiC

2.2.1 Crystal Structure

SiC is the only stable chemical compound of silicon and carbon. The crystalline structure of SiC can be considered as consisting of two sublattices with the densest packing of atoms, one of which is represented by Si atoms and the other one by C atoms and which is shifted along the main axis of symmetry by a quarter of the distance between the adjoining layers of Si atoms. The smallest distance between Si and C atoms is 1.89 Å. Each Si (or C) atom is surrounded by four C (or Si) atoms and is bound with them by directional strong tetrahedral sp^3-bonds (bond covalency is about 88%). This is why SiC has exceptional chemical and thermal stability and great hardness.

In this densest packing two layers of atoms, layer "A" and layer "B" above it, are arranged always in the same way, while the next higher or third layer of atoms can be arranged in two alternative ways. It may become similar to layer "A" in its arrangement of atoms, or it may form a new layer "C". In the latter case, a great number of various versions for a regular succession of alternating layers of the above three kinds are possible, from the simplest types of ABC ABC ABC (sphalerite) or ABABAB (vurtzite) to structures consisting of many hundreds of layers. One of the most interesting features of SiC is that various multilayer periodic structures, polytypes, are produced in the process of its crystallization (see the monograph by *Verma* and *Krishna* [2.19]).

In Fig. 2.1 the schematic arrangement of atoms in some SiC structures is shown. It is seen that in either of the sublattices of SiC multilayer polytypes there are crystallographically non-equivalent atoms which are denoted with letters "h" and "k" having numerical subscripts.

Atoms of h-type are in a hexagonal environment with respect to the closely adjacent layers whereas those of k-type are in a cubic environment. The h and k atoms of subsequent layers occupy different position, which are marked with numerals 1, 2, 3 ... Taking into acount what was said above, the crystallo-

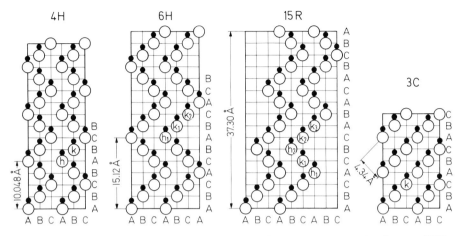

Fig. 2.1. Arrangement of Si and C atoms in plane (1120) in 4H-, 6H-, 15R-, 3C-polytypes of SiC

graphic features of different polytypes of SiC can be characterized as follows:

three-layer cubic (C) $-$ 3C-ABC ABC...$(k)_3$
four-layer hexagonal (H) $-$ 4H-ABCB...$(hk)_2$
six-layer hexagonal (H) $-$ 6H-ABCACB...$(hk_1k_2)_2$
fifteen-layer rhombohedral (R) $-$ 15R-ABCACBCABACABCB
$(h_1k_1h_2k_2k_3)_3$.

The most common polytypes are 6H, 15R and 3C-SiC. Previously it was customary to denote cubic SiC as β-SiC and all hexagonal and rhombohedral polytypes, as α-SiC.

From hk designations one can see that the number of non-equivalent positions in 3C, 4H, 6H, and 15-R SiC polytypes is equal to 1, 2, 3, and 5, respectively, and they are subdivided into two groups h and k, which are closer to each other. It is evident that when substitutional impurity atoms are included into the SiC lattice they may occupy a corresponding number of non-equivalent positions depending on the polytype.

After *Acheson* had suggested a commercial method for the synthesis of technical (abrasive) SiC [2.20], much attention was paid to the investigation of SiC properties. But not until the methods of single-crystal SiC production, whose purity approached that required for semiconductors, had been mastered [2.21–25], did it become possible to investigate the electronic properties of the material. It appeared that the electronic properties of SiC polytypes may differ considerably, while their physicochemical and mechanical properties (see Table 2.1) are very similar. The solubility of impurities in different SiC polytypes and their diffusive mobility (see Table 2.2) are also similar. Only small-size light atoms dissolve markedly in SiC when the standard methods of atom injection are used [2.34, 36, 37]. Aluminium, gallium [2.35] and most of other impurity atoms substitute for silicon in SiC, while boron, nitrogen [2.38]

Table 2.1

Properties	Unit	SiC polytype	Value	Temperature (°C)	Reference
Density	g/cm³		3.21	0	[2.26]
Thermal expansion	10^{-6} cm °C^{-1}	6H	4.2	300–500	[2.27, 28]
Hardness (Mohs)			9.2 ÷ 9.5		
Young's modulus	10^3 kg/mm²	6H	37.0	20–1000	[2.29]
Debye temperature	K		1 200		[2.30]
Specific heat	cal/gm °C		0.17	20	[2.27]
			0.27	700	[2.29]
			0.30	1200	
Thermal conductivity	W/cm °C	6H	$4.9 \ (1 \cdot 10^{17} \ \text{cm}^{-3})$	20	[2.37]
			$3.2 \ (2 \cdot 10^{18} \ \text{cm}^{-3})$		
		3C	$2.6 \ (1 \cdot 10^{16} \ \text{cm}^{-3})$	20	[2.37]
			$1.3 \ (8 \cdot 10^{20} \ \text{cm}^{-3})$		
Refractive index ($hv \sim 2$ eV)		3C	2.64	20	[2.31]
		6H ($\perp$)	2.65		
		6H ($\parallel$)	2.68		
Dielectric constant ε_∞		3C	6.52	20	[2.31]
		6H ($\perp$)	6.52		
		6H ($\parallel$)	6.70		
ε_s		3C	9.72		
		6H ($\perp$)	9.66		
		6H ($\parallel$)	10.03		

Table 2.2

Impurity atoms / Properties	Nitrogen		Oxygen		Beryllium		Boron[a]		Aluminum[a]		Gallium[a]		Scandium		Lanthanides	
	1800 °C	2450 °C	1800 °C	2300 °C	1800 °C	2300 °C	1800 °C	2300 °C	1800 °C	2300 °C	1800 °C	2300 °C	1800 °C	2300 °C	1800 °C	2300 °C
Max. solubility attained [cm⁻³]	$8 \cdot 10^{20}$	$2.6 \cdot 10^{20}$			$7 \cdot 10^{17}$	$5 \cdot 10^{19}$	$4 \cdot 10^{19}$	$2.5 \cdot 10^{20}$ $(000\bar{1})$		$7 \cdot 10^{20}$	$2.8 \cdot 10^{18}$	$7 \cdot 10^{18}$ $(000\bar{1})$	$3 \cdot 10^{17}$		$\leqslant 10^{16}$	
[cm⁻³]							$2 \cdot 10^{19}$	$1.5 \cdot 10^{20}$ (0001)		$1.1 \cdot 10^{21}$	$1.2 \cdot 10^{19}$	$1.2 \cdot 10^{19}$ (0001)				
Reference	[2.32]	[2.33]			[2.34]ᵇ		[2.34]ᵇ		[2.34]ᵇ		[2.35]		[2.36]		[2.36]	
Effective diffusion coefficient cm²/s																
Fast branch	—	—	—	—	$2 \cdot 10^{-9}$	$1 \cdot 10^{-7}$	$2 \cdot 10^{-12}$	$4 \cdot 10^{-10}$	—	—	—		$< 10^{-13}$			
Slow branch	$10^{-18} \div 10^{-17}$	$10^{-12} \div 10^{-13}$	$1.5 \cdot 10^{-16}$	$5 \cdot 10^{-13}$	$3 \cdot 10^{-12}$	$1 \cdot 10^{-9}$	$2.5 \cdot 10^{-13}$	$3 \cdot 10^{-11}$	$3 \cdot 10^{-14}$	$6 \cdot 10^{-12}$	$2.5 \cdot 10^{-14}$	$3 \cdot 10^{-12}$				
Reference							[2.34]ᵇ									

ᵃ The crystal face on which the solubility was studied is in parenthesis (doping was performed during the epitaxial growth); $(000\bar{1})$ – carbon, (0001) – silicon faces.
ᵇ Some data have been defined more accurately as compared to [2.34].

and probably beryllium substitute for carbon atoms. Small interatomic distances and high bond energy result in a low migration ability of atoms in the SiC lattice. Most impurity atoms migrate via vacancies in SiC and only beryllium, lithium and probably hydrogen can migrate via an interstitial mechanism. The results of diffusion studies are summarized in [2.34].

2.2.2 Electronic Properties

Electronic properties of silicon carbide have been studied in a considerably less detailed way than those of germanium, silicon or of the most common $A^{III} B^{V}$ compounds. This is primarily due to the lack of pure crystals (impurities below 10^{17} to 10^{16} cm^{-3}). Therefore, no determination for example, of the values of effective masses in SiC has been made so far by the most direct method of cyclotron resonance.

The electrophysical parameters of the main SiC polytypes are given in Table 2.3. The theoretically calculated data for the 3C, 2H, 6H, and 4H-SiC band structures [2.39, 40], experimental results of absorption, Raman scattering [2.56], radiative recombination and transfer phenomena were taken into account in the tabulation. The Brillouin zone for a hexagonal structure is shown in Fig. 2.2.

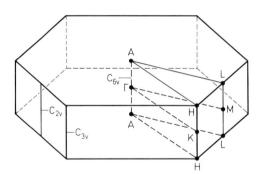

Fig. 2.2. Brillouin zone for hexagonal structure

It follows from the calculation that the absolute minimum of the conduction band in 3C-SiC is located at point X, and in the hexagonal polytypes it is on the M-L line. The calculation accuracy does not permit to choose the absolute minimum position along this line in a more accurate way. Therefore when chosing this point, one proceeds from the electrical absorption data [2.43] and from the value of the density of states effective mass in the conduction band. It is seen from Table 2.3 that *p*-type SiC has nearly similar properties in the different polytypes, whereas the *n*-type SiC polytypes differ greatly [2.47].

Table 2.3

SiC polytype / Properties	Symbol	Unit	3C-SiC	Reference	15R-SiC	Reference	6H-SiC	Reference	4H-SiC	Reference
Lattice parameter	a / c	Å	4.359		3.073 / 37.3		3.081 / 15.12		3.076 / 10.05	[2.40]
Localization of absolute max. of valence-band			Γ	[2.39] [2.40]	Γ		Γ	[2.40]	Γ	
Most probable localization of absolute min. of conduction-band			X		line M-L		M		near M-L	
Most probable number of valleys	M_C		3		3		3		6	
Exciton energy gap-0K	E_{gx}	eV	2.3	[2.41]	2.986	[2.41]	3.023	[2.41]	3.265	[2.41]
Binding energy of exciton		eV	0.014	[2.42]	0.040^a		0.078	[2.43]	0.020	[2.43]
Energy gap	E_g	eV	2.402		~3.020		3.090		3.280	
Nitrogen ionization energy	$E_D^h(N)^b$ / $E_D^k(N)$	eV	0.05	[2.55] [2.31]	0.047 / ~0.1	[2.44]	0.095 / ~0.15	[2.44]	0.033 / 0.09	[2.44] [2.45]
Aluminum ionization energy	$E_A^T(Al)^c$ / $E_A^0(Al)$	eV	0.257 / ÷0.266	[2.55] [2.31]	~0.27	[2.47]	~0.27 / 0.27 / ÷0.3	[2.46, 47] [2.45] [2.48]	~0.27 / 0.20	[2.47] [2.45]
Boron ionization energy	$E_A^T(B)$ / $E_A^0(B^*)$	eV	0.73	[2.49]	0.39 / 0.73		0.39 / 0.39 / 0.73	[2.117] [2.48]	0.39 / 0.73	

Table 2.3 (continued)

SiC polytype / Properties	Symbol	Unit	3C-SiC	Reference	15R-SiC	Reference	6H-SiC	Reference	4H-SiC	Reference
Beryllium ionization energy	E_A^T (Be)	eV					1st Level- 0.4 2nd Level- 0.6	[2.50]		
Gallium ionization energy	E_A^T (Ga) / E_A^O (Ga)	eV					0.29 / 0.35	[2.35]	0.27	[2.51]
Scandium ionization energy	E_A^T (Sc)	eV					0.24	[2.52]		
Anisotropy of conductivity	$\varrho_\parallel / \varrho_\perp$		1		1.5	[2.53]	3.7	[2.53]	0.9	[2.53]
Effective electron mass in the valley	$m^*_\perp$ / $m^*_\parallel$ / $m^c_{al.}$	in m_0 unit	0.1÷0.15		0.16-0.25 / 0.24-0.37 / 0.19-0.29		0.29-0.35 / 1.1-1.4 / 0.38-0.45		0.17-0.21 / 0.16-0.21 / 0.17-0.21	
Effective mass of density of states in conduction-band	m^*_{dc}		0.24-0.35	[2.54]	0.25-0.37	[2.44]	0.6-0.7		0.33-0.40	
Electron mobility 300 K	μ_n	$\frac{cm^2}{v \cdot s}$	>1000		500		330	[2.44]	700	[2.44]
Hole mobility 300 K	μ_p	$\frac{cm^2}{v \cdot s}$	60		60		60		60	

[a] Estimated value taking into account E_D (N) and m^* for this polytype.
[b] Superscript h refers to a group of shallow non-equivalent levels of N, k to a group of deep levels.
[c] Superscript T refers to thermal energy of ionization, O to optical ionization energy.

2.2.3 Optical Properties

The most detailed review of the optical properties of SiC is given in [2.31, 41], and that of transfer phenomena in [2.46, 47]. In [2.57], a non-standard approach is used to consider the SiC polytypes as super-periodic structures of the main polytype 3C. The principal features of the impurity absorption spectra (see Fig. 2.3) for a great number of SiC polytypes were explained on the basis of this approach. At low-temperature the absorption peak is due to the optical transistion of electrons from donor levels (nitrogen) to higher energy minima of mini-bands arising, according to the author of [2.57], from the superposition of periodic potentials. The fact that the observed absorption peaks in both polarizations are proportional to the concentration of uncompensated donor

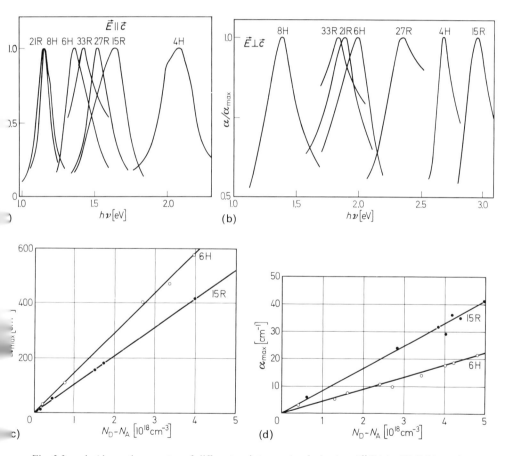

Fig. 2.3 a–d. Absorption spectra of different polytypes at polarization $E\|C$ (a); $E \perp C$ (b); and dependence of absorption coefficient on concentration of $N_D - N_A$ $E\|C$ (c); $E \perp C$ (d) in 6H- and 15R-SiC; 300 K

atoms is of great practical importance: it permits a simple and efficient optical method for selecting SiC crystals on the basis of their homogeneity and degree of doping.

The presence of absorption peaks in the visible region of the spectrum (see Fig. 2.3) causes the different coloring of doped n-SiC crystals of different polytypes.

A comparative analysis is most useful for explaining many electronic properties of the principal SiC polytypes. Although non-equivalent atoms of nitrogen and boron in 6H-SiC were first discovered by the analysis of EPR spectra [2.38], the presence of non-equivalent positions of substitutional atoms was most convincingly demonstrated when the optical properties of SiC were investigated. *Choyke* and *Patrick* discovered and studied [2.41] the short wavelength low-temperature L, which appears in different polytypes at the annihilation of the exciton bound to a neutral donor, nitrogen (see Fig. 2.4). In a general way, for the simplest 3C polytype it is similar to the L, in silicon [2.58] and diamond [2.6].

Similar spectra were studied for 3C, 2H, 4H, 6H, 15R, 21R, 33R polytypes. This L is characterized by the following: a) The number of no-phonon lines (of O, P_0, Q_0, R_0, and S_0 types) is equal to or somewhat less than the number of possible non-equivalent positions of atoms for each polytype and is in accordance with the number of non-equivalent h and k atoms; these lines are collected for each polytype into two closely packed groups (see Figs. 2.4 and 2.5) on the basis of their radiation energy. b) There is a very pronounced phonon structure in the spectrum and the principal phonons are similar for all the polytypes (see Table 2.4). c) The lowest energy binding the exciton to a neutral donor in each polytype (corresponding to "hexagonal" non-equivalent positions) counted from the exciton energy gap E_{gx} correlates well with the thermal ionization energy of nitrogen $E_D(N)$ in the corresponding polytype, amounting to 0.17 to 0.2 of the $E_D(N)$ value. d) A deeper donor level observed in [2.44, 56, 59, 60] equal to 0.09 eV for 4H, 0.15 eV for 6H and 0.1 eV for 15R is in good correlation with the higher binding energy of the exciton (corresponding to "cubic" non-equivalent positions) and the ratio of these energies is also equal to 0.2.

Proceeding from the binding energy of exciton in 3C-SiC, 10 meV, $E_D(N)$ for 3C-SiC should be 0.05 eV, which is, as we shall see later, in good agreement

Table 2.4

Phonon branch	3C	15R	6H	4H
TA	46.3	46.3	46.3	46.7
LA	79.4	78.2	77.0	76.9
TO	94.4	94.6	94.7	95.0
LO	102.8	103.7	104.2	104.0

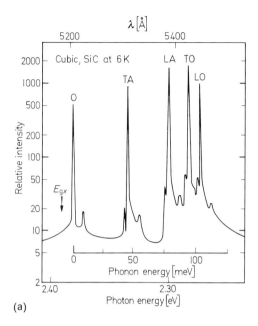

(a)

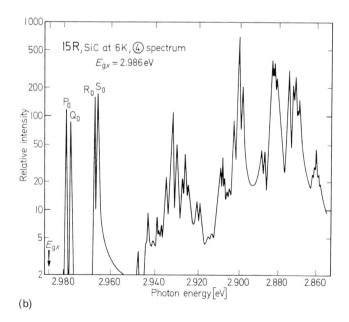

(b)

Fig. 2.4 a and b. Four-particle nitrogen-exciton luminescence spectrum at 6 K in 3 C-SiC (a) and 15 R-SiC (b)

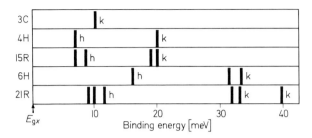

Fig. 2.5. Energy position of no-phonon lines of exciton spectra

with the results or L of DAP and the results of *Altaisky* on the thermal ionization energy of nitrogen in this polytype.

Although the radiative annihilation of excitons on the neutral atom strongly confirmed the concept of non-equivalent atoms in the SiC lattice and allowed to establish many important parameters in different SiC polytypes, it cannot be used for EL, since this type of L has low efficiency and is quenched as soon as the temperature rises. The same is true of most other types of the exciton L in SiC in particular, at defects caused by ion implantation [2.31] and at isoelectronic Ti [2.61]. Only the radiative annihilation of indirect free excitons in 3C-SiC [2.42, 62] is of practical interest for EL. It can be studied by excitation with a laser or an electron beam (CL). The radiation is accompanied by the emission of TA (46 meV), LA (78 meV), and TO (94 meV) phonons and is revealed as the corresponding three maxima. These maxima fuse into one when the temperature is raised or the excitation level is increased. The behavior of this L is similar to that in Si [2.63] and in diamond [2.64]. The most efficient L in the wider-band hexagonal polytypes of SiC is the impurity L, which is used in EL devices.

Many papers are devoted to the study of impurity L in SiC. L in the whole visible region of the spectrum in semiconducting SiC was first investigated in [2.65]. The efficient yellow, red and green EL of *pn* junctions was first obtained in [2.16]. The effect of different activators of L in SiC was first discovered for B in [2.116,66], for Al in [2.67], for Ga in [2.67,68], for Be in [2.69] and for Sc in [2.70]. The L spectra of SiC doped with different activators are shown in Fig. 2.6.

The correlation established between the efficiency of boron, yellow beryllium and scandium L and the nitrogen concentration in SiC supported the DAP mechanism of radiative recombination. This interpretation was confirmed by the line L observed on the short wavelength side of the "blue" aluminum band [2.71]. The part played by the radiative recombination through DAP (Al-N) was postulated in many studies [e.g., 2.72, 73], in which widely differing band schemes for transitions were suggested.

The type II spectrum consisting of a multitude of lines converging on the long wavelength side to a structureless maximum B_0 with phonon repetitions was observed only in [2.55] and [2.74]. This type II spectrum, which is characteristic for DAP located at different sublattices, was discovered in crystals

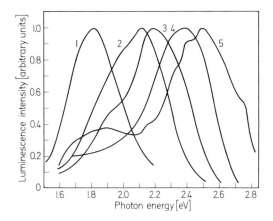

Fig. 2.6. Luminescence spectra of SiC doped with various activators:
1) 6 H-SiC (Be), 2) 6 H-SiC (B),
3) 6 H-SiC (Sc), 4) 4 H-SiC (Sc),
5) 6 H-SiC (Al)

3 C-SiC doped with N and Al, at temperatures below 2 K. When the temperature is raised above 77 K the DAP spectrum begins to weaken and a greater part of L is due to the radiative recombination of free electrons with holes bound to aluminum acceptor atoms (the so-called A-spectrum) [2.75]. The analysis of a linear spectrum allowed determination of $E_D(N) + E_A(Al) = 310$ meV. The knowledge of the energy position of a no-phonon maximum A_0 in the A-spectrum (2.153 eV) makes it possible to determine the value of $E_A(Al) = 0.257 + 0.265$ eV by the formula $E_A(Al) = E_g - E(A_0) + 2kT$. Another approach to the analysis of similar results [2.55] gave the same value of $E_A(Al)$. Hence, $E_D(N) = 0.05 \pm 0.005$ eV. These values agree very well with the results obtained by other methods [2.48, 76].

Recently a similar L was discovered in 3 C-SiC crystals doped with boron [2.49, 77]. In Figs. 2.7 to 2.10 all specific features of L associated with the presence of N and B in the 3 C-SiC lattice are illustrated.

The behavior of this type of L is similar to the L in 3 C-SiC (Al, N) in many respects, but it has the following differences: a) the line spectrum corresponds to type I and not to type II, which indicates that N and B are in the same SiC sublattice (the carbon sublattice); b) the spectrum maximum is shifted toward low energies with respect to no-phonon peaks not by one LO-phonon but by two phonons, although the phonons in both spectra are identical (LA = 69 meV and LO = 116 meV); c) the L associated with the radiative transition of a free electron to an acceptor impurity (Spectrum A) begins to appear for the SiC (B, N) specimens at lower temperatures than for SiC (Al, N) and it is quenched at much higher temperatures (the first, beginning from 400 to 500 K, and the second, from 110 to 120 K (Fig. 2.10); d) the ratio of no-phonon maxima B_0 and A_0 to phonon replicas is much less for "boron" L than for "aluminum" L. All the foregoing shows that the boron luminescence-active center is much deeper than Al, and therefore the radiative recombination through it has a greater contribution from electron-phonon interactions. Indeed, the computations of activation energies of boron and nitrogen levels made by analogy

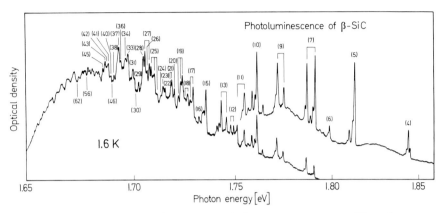

Fig. 2.7. Line spectrum of DAP B-N (type 1) in 3C-SiC, $T = 1, 6\,\mathrm{K}$ ($N_B = 1 \cdot 10^{18}\,\mathrm{cm}^{-3}$)

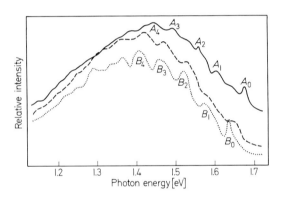

Fig. 2.8. Luminescence spectra of DAP B-N of specimens with different doping levels (3C-SiC): solid line 77K, dotted line 4.2K, $N_B = 1 \cdot 10^{18}\,\mathrm{cm}^{-3}$; broken line 77K, $N_B = 2.5 \cdot 10^{17}\,\mathrm{cm}^{-3}$

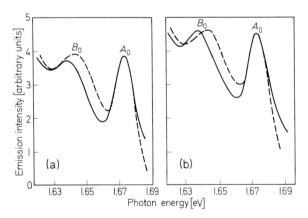

Fig. 2.9a and b. Dependence of luminescence spectra of DAP B-N on B and N concentration (a) and excitation level (b) (3C-SiC): (a) solid line $N_B = 4.5 \cdot 10^{17}\,\mathrm{cm}^{-3}$, broken line $N_B = 1 \cdot 10^{18}\,\mathrm{cm}^{-3}$, (b) the relative excitation rates are 1 and 0.02 for the broken and the solid lines, respectively

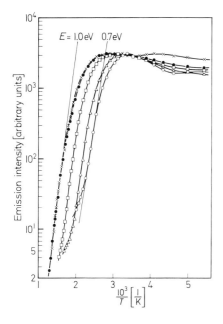

Fig. 2.10. Temperature quenching of L in 3C-SiC(B, N)

with the computation of the aluminum spectrum give the values of $E_A(B^x) = 0.73(5)$ eV and $E_D(N) = 0.05(5)$ eV.

The analysis of the [2.78] data, taking into consideration the values of E_g obtained for the main SiC polytypes, permitted establishing that values of $E_A(B^x)$ for all polytypes are really very close to the value of $E_A(B^x)$ for 3C-SiC (see Table 2.5).

It is not clear yet why the $E_A(B^x)$ value obtained from the DAP spectrum should vary from the value of $E_A(B) = 0.39$ eV obtained from an analysis of transfer phenomena and of electrooptical absorption [2.48].

The spectra of impurity L of higher order polytypes are complicated essentially by the availability of impurity levels of different energies. But the advance in understanding the nature of L spectra of 3C-SiC and that the non-equivalent impurity levels are grouped into two series (h and k) according to the ionization energies in all polytypes helped to explain the fact noted by many investigators that two series of bands [2.79, 80] appear in aluminum L in 4H-, 6H-, and 15R-SiC polytypes. This is evident in Fig. 2.11.

The no-phonon maxima of L, B_0 and C_0, are 50 meV apart in both polytypes. This separation corresponds to the difference in the activation energies of a more shallow and a deeper donor level (see Table 2.3). Thus, B_0 and C_0 peaks are equivalent in their nature to the B_0 L peak in 3C-SiC (Al, N) and they correspond to the radiative recombination through remote DAP, Al-N, shallow for B_0 and deep for C_0. The radiative transition of the A_0 type begins to appear at temperatures above 105 K in 4H-SiC [2.81]. The high energy of nitrogen ionization in 6H-SiC and the quenching of L at temperatures above

Table 2.5
E_{max} is the energy of boron L maximum at 300 K
$E_g - E_{max} = E_A(B^a) + 2LO = 0.73$ eV $+ (0.2 - 0.22)$ eV $= 0.93 - 0.95$ eV

Polytype	E_g eV	E_{max} eV	$E_g - E_{max}$ eV
4H	3.26	2.317	0.94
6H	3.08	2.137	0.94
15R	2.99	2.049	0.94
3C	2.38	1.42[a]	0.96

[a] According to the results of [2.49, 77]

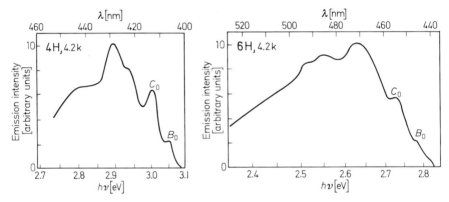

Fig. 2.11. Luminescence spectra of DAP Al-N in 6H- and 4H-SiC polytypes

100 K almost completely exclude the possibility of observing this type of L in 6H-SiC. The analysis of these spectra gives $E_A(Al)$ in 4H-SiC equal to 0.20 eV and to 0.26 eV in 6H-SiC. The "gallium" spectrum also has a similar system of lines, but, due to a greater value of $E_A(Ga)$, it is shifted toward lower energies [2.35, 51]. In the "boron" spectrum, on the contrary, the L of free electrons at neutral boron atoms is observed more easily, especially at higher temperatures. Phonon replicas in all three types of L are very close together (but for the boron L the maximum is at 2 LO, and for Al and Ga- at 1 LO phonon). The ratio of the no-phonon peak to the L at the spectral maximum diminishes in the series of Al-Ga-B. The spectra of boron, gallium and aluminium luminescence in 4H-SiC at 77 K are given for comparison in Fig. 2.12 (the radiative transition through a shallower donor level is shown).

As seen in Fig. 2.12, the L of SiC doped with gallium approaches the L of SiC doped with aluminum. The studies of the L SiC doped with Be and Sc showed, on the contrary, that in their behavior these L types are similar to the boron L.

There are the following essential features of the boron L (as well as of the yellow beryllium and scandium L): a) a rather strong brightening with increasing temperature in the 150 to 400 K range (see Fig. 2.13); b) a different character of the brightness dependence on excitation intensity in the different temperature ranges—sublinear in the region of brightening and superlinear in the region of temperature quenching [2.52, 82, 83] (see Figs. 2.14 to 2.16); c) an increased efficiency of L in the n-type material, which markedly distinguished these types of L from those of aluminum and gallium.

The elucidation of the nature of these properties and the sublinearity of L as a function of the excitation level, in particular, is of great practical importance, since at low excitation levels the internal efficiency of the boron L at 300 K can reach 40% [2.84] and the efficiency of scandium and beryllium L is not far behind that of boron L. Recently, data have been obtained suggesting that uncontrolled defects (of the trapping type) may be responsible for the sublinearity and other properties of L in SiC. Thus, under certain conditions, the injection of boron into a SiC crystal doped with Sc may result in the

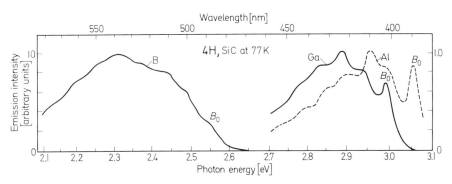

Fig. 2.12. Luminescence spectra of 4H-SiC doped with B, Ga, Al

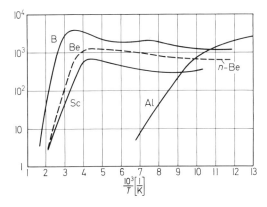

Fig. 2.13. Temperature dependence of L of 6H-SiC doped with different activators

complete quenching of the scandium L and in the brightening of the boron L, while at the same time, leaving unchanged the dependence of CL on the excitation intensity (see Fig. 2.17) [2.85].

The essential role of the traps in the specific character of the boron L was pointed out in [2.86, 87]. Such traps were discovered when studying the photoconductivity of SiC in [2.88]. Numerous results indicate that the nature of traps appears to be identical for the three types of L. In Fig. 2.18 the most common schemes of radiative transitions through impurity atoms in 3C-, 6H-, and 4H-SiC are shown. In this case, the role played by competing radiative transitions (in the IR-region in particular) and non-radiative recombination paths through uncontrolled defect centers was not taken into consideration.

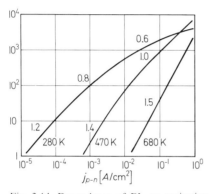

Fig. 2.14. Dependence of EL on excitation level at different temperatures

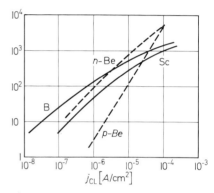

Fig. 2.15. Dependence of CL efficiency on excitation level in SiC doped with B, Be, Sc

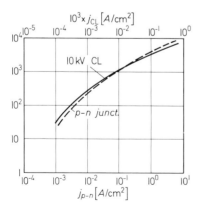

Fig. 2.16. Comparison of EL- and CL dependences on excitation level

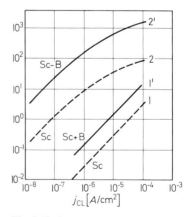

Fig. 2.17. Comparison of CL dependence on excitation level in 6H-SiC (Sc) specimens before B-diffusion into them (1,2) and after diffusion (1',2')

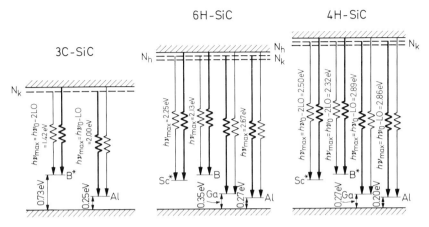

Fig. 2.18. Schemes of fundamental radiative transitions in 3C-, 6H-, 4H-SiC manifesting themselves in electroluminescence

2.3 Material Preparation

For the effective utilization of impurity L in EL devices, it is necessary to produce high quality defect-free crystals of SiC (6H- and 4H polytypes in particular) and to dope them in a controlled way.

Growing SiC crystals with the semiconducting degree of purity is a complicated technical task. It follows from the phase diagram of Si-C binary system (Fig. 2.19) that the main difficulty in growing the crystals is caused by the absence of a liquid phase of SiC at atmospheric pressure and by the high temperature of the process.

SiC crystals of semiconducting purity can be produced by growth from solution, by thermal dissociation and reduction of compounds containing Si and C or by sublimation.

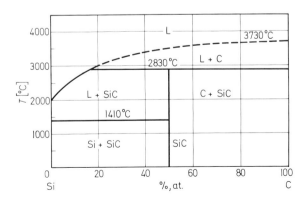

Fig. 2.19. Phase diagram of Si-C system

2.3.1 Crystal Growth from Solution

The crystallization of SiC from solution avoids the main difficulties encountered in growing SiC from the melt at the necessary high pressures, and permits lowering of the temperature of crystal growth. Si is an ideal solvent for SiC for growing crystals of semiconducting purity. Unfortunately, the very low solubility of C in Si prevents the growth of SiC crystals at rates acceptable for practical applications. The solubility of C in the melt can be increased by using a number of pure metals (Cr, rare-earth elements) which have a low solubility in SiC [2.89–91]. As follows from Fig. 2.20 the solubility of SiC in rare-earth metals at 1800 °C lies in the range of 40 to 65 mol%.

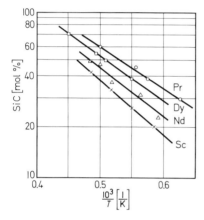

Fig. 2.20. Dependence of silicon carbide solubility in rare-earth metals on temperature

The methods of directional crystallization, zone melting with temperature gradient, etc., can be used for growing SiC from the solution. The use of rare-earth elements as a solvent of SiC permits growing of SiC crystals at rates of 1 mm/h and higher. The stable growth of a certain polytype is an important feature of crystal growth from solution in rare-earth elements. The stabilization of the growth of a rather rare polytype, 4H-SiC, was observed when crystals were grown from an Sc solution [2.92].

2.3.2 Crystal Growth by Thermal Decomposition

Single-crystal platelets of SiC up to 6 to 7 mm across can be grown by the thermal decomposition of organic silicon compounds, such as methyl-trichloro-silane (CH_3SiCl_3) dimethyl-chlorosilane ((CH_3)$_2SiCl_2$), methyl-dichlorosilane (CH_3SiHCl_2) and other compounds [2.93, 94]. The growth of 3C-SiC crystals with this technique is being developed to increase the size of the crystal, and to improve its perfection and purity.

2.3.3 Crystal Growth by Sublimation

At present, the most common method of growing semiconductor grade SiC crystals is the sublimation technique first suggested by *Lely* in 1956 [2.21]. *n*- and *p*-type crystals of silicon carbide up to 30 mm and more across are grown by this method. In the sublimation technique, the crystals are grown in cylindrical graphite crucibles (Fig. 2.21).

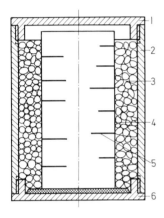

Fig. 2.21. Cross section of crucible used for growing crystals by sublimation: 1, 2, 6) Graphite crucible; 3) Powdered SiC; 4) Graphite cylinder with holes; 5) Growing crystals

The crystal growth zone consists of an apertured thin-walled graphite cylinder. Powdered SiC synthesized from pure Si and C is used as a starting material. The crucible is heated in a cylindrical resistance oven using a graphite heater or in an electronic oven. Crystals are grown in an inert gas atmosphere (most frequently argon) at 2400 to 2600 °C. The nucleation of crystals takes place in the growth zone, mainly in the holes of the graphite tube. Thin hexagonal platelets are obtained.

SiC crystals are also grown by a modified sublimation technique, e.g., when pure Si is used as a starting material and C is supplied by the graphite accessories of the oven [2.95]. The advantage of this method is the possibility of producing crystals of a higher purity, as well as the simplicity of controlling the supersaturation of SiC vapor in the growth zone.

Doping SiC with donor and acceptor impurities during the growth process is realized by introducing N or some volatile compounds, such as $Al(C_2H_5)_3$, BF_3, etc., into the growth atmosphere. The donor concentration in SiC crystals increases with the partial pressure (P_{N_2}) of N in the oven atmosphere as $N_D = 6.5 \times 10^{18} \sqrt{P_{N_2}}$ and it can vary from 10^{16} to 2.5×10^{20} cm^{-3} [2.96].

n-SiC crystals doped with N are used to produce different types of EL devices.

2.4 EL Devices

Impurities are introduced into SiC crystals by diffusion or by ion implantation to generate *pn* junctions. The principal parameters for the diffusion of various impurities into SiC are given in [2.34].

The SiC electroluminescent devices are very interesting because of their exceptionally high stability, long operational life and their resistance to current overloads, heating and other external effects. One can obtain SiC EL devices emitting any color in the visible region of the spectrum, up to the violet. The color of SiC luminescence can be controlled in EL devices both by using crystals of a different polytype and by introducing various activators of the luminescence. The photon energy of the radiated light can be changed by 0.8 eV (see Table 2.3) just due to the differences in the energy gaps of different SiC polytypes without changing the methods of device fabrication. As follows from Fig. 2.6 light-emitting devices with red, yellow, green and blue luminescence can be made with only one SiC polytype, e.g., 6H-SiC, by introducing into SiC Be, B, Sc, and Al, respectively [2.97].

2.4.1 Device Fabrication

EL devices can be manufactured by different methods of semiconducting technology: by alloying, epitaxy during crystal growth, diffusion and ion implantation [2.16, 66, 98–102]. Thus, SiC EL *pn* junctions with red luminescence can be obtained by alloying Si in an N-containing atmosphere onto a *p*-SiC crystal homogeneously doped with Be (this doping having been obtained during the growth of the crystals by the sublimation technique or from the solutions of Be-SiC, Be-Cr-SiC [2.103]).

The gaseous-phase or the liquid-phase epitaxy of *p*-SiC(Be) layers on *n*-SiC substrates results in the formation of EL *pn* junction in the substrate due to the fast diffusion of Be.

Homogeneously luminescent *pn* junctions can be formed in the basal plane of *p*-SiC(Be) crystals if a small amount of N is introduced into the growth atmosphere during the final stage of the growth process by the sublimation technique.

Efficient EL *pn* junctions with green luminescence are produced by liquid or gaseous epitaxy of *n*-SiC layers doped with N and with the luminescence activator, Sc. Liquid epitaxy is used for growing an epitaxial layer of *n*-SiC(Sc) from Sc solution on *p*-SiC substrates doped with Al. EL *pn* junctions are also formed by successive gaseous epitaxy of *n*-SiC(Sc, N) and *p*-SiC layers on *n*-SiC(N) substrates.

Epitaxy is produced by evaporation of polycrystalline SiC in an inert atmosphere with a subsequent condensation of vapors on substrates which are supercooled with respect to the vapor. When the epitaxial layer is to be doped with Sc, the latter is introduced into the source material during its synthesis. The concentration of Sc in the epitaxial layers amounts to (2 to 3) ×

10^{17} cm^{-3}, i.e., close to the solubility limit of Sc in SiC (see Table 2.2). The subsequent growth of the p-SiC epitaxial layer doped with aluminum having the concentration of $N_{Al} = 5 \times 10^{19}$ cm^{-3} leads to the formation of p^+-SiC(Al) − n-SiC(Sc, N) structures. During a short-term epitaxy (~ 10 min) of the p^+-SiC(Al) layer at 2200 °C the pn junction is formed as a result of the Al diffusion into the SiC(Sc, N). The 0.05 μ diffusion improves the quality of the pn junction.

EL pn junctions with blue luminescence are created during the Al-doping of silicon carbide by the methods of alloying, diffusion and epitaxy. pn junctions are produced by the diffusion of pure Al from the vapor phase at 2000 to 2400 °C for 4 to 10 hours. p-SiC layers, 100 to 200 μm thick, homogeneously doped with Al are grown on n-SiC substrates during the epitaxial growth [2.104].

The simplicity of manufacturing SiC(B) EL devices with various configurations of luminescent areas and high efficiency of EL are the reasons for their more extensive use. SiC(B) EL pn junctions with yellow luminescence are formed on the 6H polytype principally by the diffusion method. The boron diffusion into n-SiC crystals is produced at 1800 to 2300 °C for 2 to 60 min. A rather great temperature dependence of B solubility in SiC [2.34] and a high activation energy permit formation of controllably resistive p-region near the pn junction (the resistivity of a p-SiC(B) layer amounts to 10^{13} ohm·cm). This characteristic feature of the p-layer combined with its small thickness (amounting to a few microns) provides for the isolation of luminescent regions produced in these single-crystal EL devices. The configuration of the luminescent region reproduced faithfully the configuration of the contact electrode on the p-side of the junction. The spreading of the current away from the contact edge does not exceed 8 to 10 μm [2.105].

Efficient EL pn junctions can be also produced by independently doping SiC with Al and boron [2.106]. Taking into consideration that in SiC E_A(Al) $< E_D$(B), the Al diffusion causes the formation of a pn junction with a more efficient injection of holes, and the subsequent B diffusion forms an efficient luminescent region. The devices created with such pn junctions are able to function in a wider temperature range, starting at 77 K. The creation of multi-element devices based on such pn junctions, however, represents a more complicated task and requires the creation of mesa-structures [2.107]. The main characteristics of EL pn junctions produced in 6H-SiC doped with various activators of luminescence are presented in Table 2.6.

Table 2.6

No.	Properties[a]	Unit	Boron pn junction		Aluminium pn junction	Scandium pn junction	Beryllium pn junction
			average	superior			
1	EL brightness at $j = 0.2$ A/cm²	candle/m²	50	200	–	30	20
	$j = 0.5$ A/cm²		90	350	10	180	50
2	Maximum EL brightness[b]	candle/m²	700	1500	500	1000	300
3	Forward voltage drop	V	2.3 – 3	2.2 – 2.4	2.4 – 3.5	2.3 – 3	4 – 6
4	Temperature range	°C	–40 – +200	–100 – +200	–160 – +50	–100 – +100	–40 – +70
5	Color of luminescence		yellow		blue	green	red
6	Maximum of EL spectrum	nm (eV)	590 – 580 (2.05 – 2.15)		460 (2.66)	560 (2.2)	680 (1.82)
7	Half-width of EL spectrum	eV	0.4 – 0.5		0.3	0.45	0.5
8	Speed of operation	s	$(1 – 3) \, 10^{-5}$	10^{-6}	10^{-8}	$(1 – 8) \, 10^{-5}$	$(1 – 6) \, 10^{-6}$
9	Specific capacitance of pn junction (at zero bias)	μF/cm²		0.05 – 0.1	0.1 – 0.15	0.06 – 0.1	0.08 – 0.1
10	Max. area of pn junction	cm²	1	2	$2 \cdot 10^{-2}$	1	$2 \cdot 10^{-2}$

[a] The values, except for item 4, are given for $T = 20\,°C$. The values of brightness are given for flat pn junction surface emitter without antireflection coating.

[b] Obtained under continuous excitation without any special cooling.

2.4.2 Potential Applications

The following EL devices are made using SiC *pn* junctions with various activators of luminescence: light-emitting diodes, pulsed light sources, digital and
symbol indicators, devices with controlled geometry of the light emitting area,
multi-element scales for data recording on photosensitive material, etc.
[2.100, 105]. The advantages of the above-mentioned devices, aside from their
high stability, include low values of operating current, which simplifies their
matching to microcircuits. The group of SiC devices which have a variable
geometry of the light emitting area, changing under the action of an applied
electric field, is of a great interest. The simplest model of such a device is a
pn junction obtained by B-diffusion, in which the *n*-region is an equipotential
while the *p*-region, produced with specified values of resistivity, is provided
with two electrodes and acts as a voltage divider (Fig. 2.22).

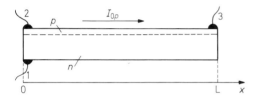

Fig. 2.22. Schematic diagram of a
device with controlled geometry of
the light-emitting area

When the current passes through contacts 1 and 2 the *pn* junction lumines-
ces homogeneously over its whole area. When the current passes through
contacts 2 and 3 simultaneously, a potential gradient is created in the voltage
divider. In this case, one part of the *pn* junction may be under-biased, i.e.,
insufficiently biased for passing the current through that position of *pn* junction. A smooth variation of the control current leads to a corresponding shift
of the boundary between the luminescent and non-luminescent parts of the
crystal. The current sensitivity of such a device is 1.5 mm/mA with the active
element dimensions equal to 8×0.2 mm^2 and the initial current density of
0.05 A/cm^2. The control current consumption is 3 to 10 mA. The light and the
dark field contrast in the crystal is sufficient for resolving the boundary position with an accuracy of not less than 10%. Similar devices can be used as
signal level indicators in integrated circuits. The analog indicating device,
using multi-element SiC light-emitting diodes converts the input control
voltage into a proportional change in the length of a luminescent strip or into
a light spot [2.108]. EL *pn* junctions with a controlled luminescence boundary
are used for recording on photographic film [2.109].

Multi-element digital EL display, which permit synthesis of any number,
are used for digital information recording on light-sensitive materials. Multi-
element *pn* junction light-emitters are used for the photographic recording of
information, for example as a binary code (see Fig. 2.23).

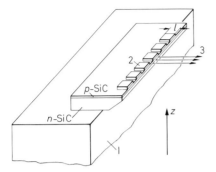

Fig. 2.23. Schematic diagram of multi-element light-emitting diodes with light output from edge of *pn* junction

2.4.3 LED Characteristics

One of the most important characteristics of the device for photorecording information is the rate of EL brightening and decay. The rise time of brightening of the average EL SiC(B) light-emitting diodes is 10 to 30 μs, the time of decay is 0.1 μs. The rise time of the EL brightening is reduced to a few μs when the current density through the *pn* junction is increased to 100 A/cm^2. The EL rise and decay times in Be-doped light-emitting diodes is ~ 0.25 μs, and that of Sc-doped *pn* junctions is larger by one order of magnitude. The relaxation time of Al-doped light-emitting diodes does not exceed 10^{-8} s, which permits the fabrication of high-speed light-emitting diodes.

Typical volt-ampere, brightness-voltage and capacitance-voltage characteristics of *pn* junctions made by diffusion with blue and yellow luminescence are given in Fig. 2.24.

It should be noted that when explaining experimental characteristics of SiC diodes, one must take into consideration high-resistivity layers which are

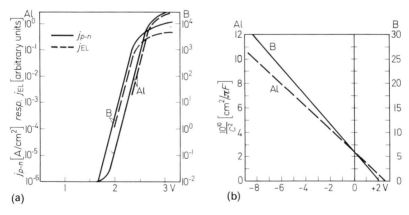

Fig. 2.24 a and b. Typical current-voltage and brightness-voltage (a) and capacitance-voltage (b) characteristics of EL in SiC(B) and SiC(Al) *pn* junctions

often formed close to a *pn* junction, defects which short-circuit the junction, as well as the presence of various traps.

Very often, microparticles of carbon and dislocations cause defects, which appear on the volt-ampere characteristics at reverse and low forward biases. The formation of high-resistance layers near a *pn* junction is associated either with deep electrically active centers in the initial material [2.110] or with the generation of deep levels during the diffusion [2.106]. Volt-ampere characteristics of SiC *pn* junctions are observed in [2.111, 112] in accordance with the theory of $p^+ - n - n^+$ structures at high levels of injection. This theory accounts for the power dependence of current on voltage ($I = (U - U_0)^m$) at high current densities. At low forward bias (up to 1.5 V) "redundant" currents are often observed which are explained in [2.113] by tunneling of carriers through intermediate states in the forbidden band.

As we have already mentioned, current-brightness characteristics of EL SiC *pn* junctions have a sublinear dependence similar to the current-brightness characteristics of CL. But the efficiency of EL appears to be lower than that of CL by 1 or 2 orders of magnitude, when they are normalized to an identical level of excitation (see Fig. 2.16); this is associated with the low level of hole injection into a luminescently active *n*-region near the *pn* junction [2.114]. The low level of hole injection is associated with the fact that the activation energy of acceptors in SiC is much larger than the activation energy of the donor (nitrogen). From the above, it follows that an increase in EL efficiency in SiC *pn* junctions may be achieved either by creating a perfect $p^+ - n$ structure with efficient hole injection into a luminescently active *n*-layer, or else by creating an efficiently luminescent *p*-SiC. Diffusion-epitaxial technology and ion implantation appear to be the most suitable for creating $p^+ - n$ structures which are efficient hole injectors.

The increase of EL efficiency in SiC *pn* junctions can be brought about also by diminishing the sublinearity of current-brightness characteristics. In Fig. 2.25 we can see how the parameters of EL in SiC(B) *pn* junction are inferior compared to, for example, GaP(ZnO) light-emitting diodes when the level of excitation is raised [2.106].

The example of green EL in 3C-SiC *pn* junctions also demonstrated how substantial it is for light efficiency and for producing high brightness to have a linear or super-linear current-brightness characteristic. In such EL *pn* junctions the radiative annihilation of indirect free excitons is most efficient at high levels of excitation above room temperature. Brightness of tens of thousands candles/m^2 has been already achieved in such *pn* junctions [2.115].

As follows from the consideration of specific features of boron, beryllium and scandium L in SiC, obtaining a more perfect quality of SiC with a much lower level of minority carrier traps is a necessary condition for the reduction of sublinearity. Studies made in recent years with the aim of improving the sublimation method of silicon carbide crystal growth, the application of this method to the production of epitaxial layers, the possibility of growing epitaxial layers and crystals of a specified polytype of SiC (especially 4H-SiC

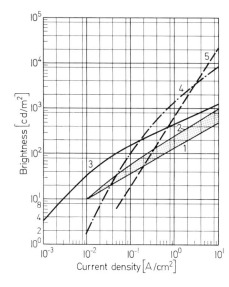

Fig. 2.25. Current-brigthness characteristics of EL in: 4,5) GaP(Zn-O) and GaP(N) *pn* junctions; 1,2) SiC(B); 3) best SiC(B) *pn* junction

from solution using rare-earth metals as solvents) — all this makes possible a successful realization of high — quality silicon carbide for the production of different types of highly stable EL devices and opens up new ways for improving their efficiency.

Acknowledgment. The authors would like to thank Drs *V. S. Vavilov, E. A. Konorova, Yu. M. Altaisky, A. P. Andreev, E. E. Violin, A. V. Petrov, G. B. Dubrovski, E. N. Mokhov*, and *G. A. Lomakina* for providing us with some new preprints. We would also like to thank *M. N. Vanina* for the translation.

References

2.1   J. R. Haynes, M. Lax, W. Flood: J. Phys. Chem. Solids **8**, 392 (1959)
2.2   C. Benoit a la Guillaume, Y. Cernogora: *Proc. Symp. Radiative Recombination* (Dunod et Cie., Paris 1964) p. 121
2.3   Ya. I. Pokrovskii, K. N. Svistunova: Soviet Phys. Solid State **6**, 13 (1964)
2.4   V. S. Vavilov, O. G. Koshelev, Yu. P. Koval, Ya. G. Klyava: Soviet Phys. Solid State **8**, 2110 (1967)
2.5   P. J. Dean: Phys. Rev. **139**, A 588 (1965)
2.6   P. J. Dean, E. C. Lightowlers, D. R. Wight: Phys. Rev. **140**, A 352 (1965)
2.7   V. S. Vavilov, G. P. Golubev, E. A. Konorova, E. L. Nolle, V. F. Sergienko: Soviet Phys. Solid State **8**, 1210 (1966)
2.8   P. S. Walsh, E. C. Lightowlers, A. T. Collins: J. Luminesc. **4**, 369 (1971)
2.9   P. Denam, E. C. Lightowlers, P. J. Dean: Phys. Rev. **161**, 762 (1967)
2.10   V. S. Vavilov, V. V. Galkin, M. I. Guseva, L. M. Ershova, K. V. Kiseleva, E. A. Konorova, V. V. Krasnopevzev, Yu. V. Milutin, V. F. Sergienko: *Ion Implantation in Diamond* (FIAN USSR, Moscow 1969)
2.11   J. R. Prior, F. C. Champion: Proc. Phys. Soc. **80**, 849 (1962)

2.12   V. V. Palogrudov, Yu. G. Penzin, E. E. Penzina: Opt. i Spektroskopiya **17**, 230 (1964)
2.13   O. V. Losev: Telegraphia i telephonia **18**, 45 (1923) and **26**, 403 (1924)
2.14   H. C. Chang, L. J. Kroko: AIEE Paper **59** (1957)
2.15   C. A. A. J. Greebe, W. F. Knippenberg: Phil. Res. Rept. **15**, 110 (1960)
2.16   Yu. S. Blank, Yu. A. Vodakov, A. A. Mostovskii: Soviet Phys. Solid State **5**, 1621 (1964)
2.17   E. E. Violin, G. F. Kholuyanov: Soviet Phys. Solid State **6**, 465 (1964)
2.18   G. F. Lymar, Yu. V. Polikanov: In *Silicon Carbide* (Naukova Dumka, Kiev 1966) p. 232
2.19   A. R. Verma, P. Krishna: *Polymorphism and Polytypism in Crystals* (John Wiley and Sons, New York 1966)
2.20   E. G. Acheson: Brit. Pat. **17**, 911 (1892); J. Franklin Inst. **136**, 194 (1893)
2.21   A. Lely: Ber. Deutsch. Keram. Ges. **32**, 229 (1955)
2.22   D. R. Hamilton: J. Electrochem. Soc. **105**, 735 (1958)
2.23   W. F. Knippenberg: Phil. Res. Rept. **18**, 161 (1963)
2.24   S. A. Nedopekina, M. V. Reifman, I. A. Surovtseva: In *Silicon Carbide* (Naukova Dumka, Kiev 1966) p. 207
2.25   Yu. M. Tairov: Rost Kristallov (USSA, Moscow) **6**, 199 (1965)
2.26   A. H. Gomes de Mesqita: Acta Cryst. **23**, 610 (1967)
2.27   E. L. Kern, E. Elevzleywine: Semicond. Proc. **8**, 28 (1965)
2.28   A. Taylor, R. M. Jones: In *Silicon Carbide — A High Temperature Semiconductor*, ed. by J. R. O'Conner, J. Smiltens (Pergamon Press, London 1960) p. 147
2.29   P. T. B. Shaffer: *Plenum Press Handbooks of High Temperature Materials*, No. 1 Materials Index (Plenum Press, New York 1964)
2.30   G. A. Slack: J. Appl. Phys. **35**, 3460 (1964)
2.31   W. J. Choyke, L. Patrick: In *Silicon Carbide–1973*, ed. by R. C. Marshall, J. W. Faust, C. E. Ryan (University of South Carolina Press, Columbia, South Carolina, 1974) p. 266–283
2.32   A. A. Pletyshkin, L. M. Ivanova: *Trans. III All-Union Conference on SiC* (Moscow 1970) p. 43
2.33   R. I. Scace, G. A. Slack: J. Chem. Phys. **42**, 805 (1965)
2.34   Yu. A. Vodakov, E. N. Mokhov: In *Silicon Carbide–1973*, ed. by R. C. Marshall, J. W. Faust, C. E. Ryan (University of South Carolina Press, Columbia, South Carolina, 1974) p. 508
2.35   Yu. A. Vodakov, G. A. Lomakina, E. N. Mokhov, E. I. Radovanova, V. I. Sokolov, M. M. Ysmanova, G. F. Yuldashev, B. S. Machmudov, Phys. Stat. Sol. A **35**, 37 (1976)
2.36   G. F. Yuldashev, M. N. Usmanova, Yu. A. Vodakov, D. M. Tadjiev: *Applied Nuclear Physics* (FIAN, Tashkent 1973) p. 12
2.37   S. H. Hagen, A. W. C. van Kemenade: J. Luminesc. **6**, 131 (1973) and **9**, 9 (1974)
2.38   H. H. Woodbury, G. W. Ludwig: Phys. Rev. **124**, 1083 (1961)
2.39   F. Herman, J. P. Van Dyke, R. L. Kortum: Mater. Res. Bull. **4**, 8167 (1965)
2.40   H. C. Junginger, W. van Haeringen: Phys. Stat. Sol. **37**, 709 (1970)
2.41   W. J. Choyke: Mater. Res. Bull. **4**, S 141 (1969)
2.42   D. S. Nedzvetskii, B. V. Novikov, N. K. Prokof'eva, M. B. Reifman: Soviet Phys. Semicond. **2**, 914 (1969)
2.43   V. I. Sankin, G. B. Dubrovskii: Soviet Phys. Solid State **17**, 1191 (1975)
2.44   G. A. Lomakina, G. F. Kholuyanov, R. G. Verenehikova, E. H. Mokhov, Yu. A. Vodakov: Fiz. Tekh. Poluprov **6**, 1133 (1972)
2.45   S. H. Hagen, A. W. C. van Kemenade, G. A. W. van der Does de Bye: J. Luminesc. **8**, 18 (1973)
2.46   H. J. Van Deal: Phil. Res. Rep. **20**, Suppl. 3 (1965)
2.47   G. A. Lomakina: In *Silicon Carbide–1973*, ed. by R. C. Marshall, J. W. Faust, C. E. Ryan (University of South Carolina, Columbia, South Carolina, 1974) p. 520
2.48   D. V. Dem'yanchik, G. A. Karatynina, E. A. Konorova: Soviet Phys. Semicond. **9**, 900 (1975)
2.49   S. Yamada, H. Kuwabera: In *Silicon Carbide–1973*, ed. by R. C. Marshall, J. W. Faust, C. E. Ryan (University of South Carolina, Columbia, South Carolina, 1974) p. 305
2.50   Yu. P. Maslokovets, E. N. Mokhov, Yu. A. Vodakov, G. A. Lomakina: Soviet Phys. Solid State **10**, 634 (1968)
2.51   A. Suzuki, H. Matsunami, T. Tanaka: Japan. J. Appl. Phys. **14**, 891 (1975)
2.52   Yu. M. Tairov, I. I. Khlebnikov, V. F. Tsvetkov: Phys. Stat. Sol. **25**, 349 (1974)

2.53   G. A. Lomakina, Yu. A. Vodakov: Soviet Phys. Solid State **15**, 83 (1973)
2.54   W. E. Nelson, F. A. Halden, A. Rosengren: J. Appl. Phys. **37**, 333 (1966)
2.55   N. N. Long, D. S. Nedzvetskii, N. K. Prokof'eva, M. V. Reifman: Opt. Spektroskopiya **29**, 727 (1970)
2.56   P. H. Colwell, M. V. Klein: Bull. Am. Phys. So. **16**, 395 (1971)
2.57   G. B. Dubrovskii: In *Silicon Carbide–1973*, ed. by R. C. Marshall, J. W. Faust, C. E. Ryan (University of South Carolina, Columbia, South Carolina, 1974) p. 207
2.58   J. R. Haynes: Phys. Rev. Lett. **4**, 361 (1960)
2.59   P. J. Dean, R. L. Hartman: Phys. Rev. (B) **5**, 4911 (1972)
2.60   G. B. Dubrovskii, E. I. Radovanova: Phys. Stat. Sol. **48**, 875 (1971)
2.61   A. W. C. Van Kemenade, S. H. Hagen: Solid State Commun. **14**, 133 (1974)
2.62   I. I. Geiezy, A. A. Nesteroy: In *Silicon Carbide–1973*, ed. by R. C. Marshall, J. W. Faust, C. E. Ryan (University of South Carolina, Columbia, South Carolina, 1974) p. 313
2.63   V. S. Vavilov, E. L. Nolle: Soviet Phys. Semicond. **2**, 616 (1968)
2.64   V. S. Vavilov, G. P. Golubev, E. A. Konopova, E. L. Nolle, V. F. Sergeiev: Soviet Phys. Solid State **8**, 1210 (1962)
2.65   I. A. Lely, F. A. Krolger: In *Semiconductors and Phosphors*, ed. by M. Schoene, H. Welker (Interscience, New York 1958) p. 525
2.66   E. E. Violin, G. F. Kholuyanov: Soviet Phys. Solid State **6**, 1331 (1964)
2.67   A. A. Addamiano: J. Electrochem. Soc. **113**, 134 (1966)
2.68   G. F. Kholuyanov: Soviet Phys. Solid State **7**, 2620 (1966)
2.69   A. A. Kal'nin, V. V. Pasynkov, Yu. M. Tairov, D. A. Yas'kov: Soviet Phys. Solid State **8**, 2381 (1967)
2.70   Kh. Vakhner, Yu. M. Tairov: Soviet Phys. Solid State **11**, 1972 (1970)
2.71   V. J. Sokolov, V. V. Makarov, E. N. Mokhov: Soviet Phys. Solid State **12**, 229 (1970)
2.72   V. I. Pavlichenko, I. V. Ryzhikov, Yu. M. Suleimanov, Yu. M. Sivaidak: Soviet Phys. Solid State **10**, 2205 (1969)
2.73   I. S. Gorban, G. N. Mishinova, Yu. M. Suleimanov: Fiz. Tverd. Tela **7**, 3694 (1965)
2.74   W. J. Choyke, L. Patrick: Phys. Rev. B **2**, 4959 (1970)
2.75   G. Zanmarchi: J. Phys. Chem. Solids **29**, 1727 (1968)
2.76   O. V. Vakulenko, O. A. Govorova: Soviet Phys. Solid State **13**, 520 (1971)
2.77   H. Kuwabara, S. Yamada: Phys. Stat. Sol. **30a**, 739 (1975)
2.78   A. Addamiano: J. Electrochem. Soc. **111**, 1294 (1964)
2.79   E. E. Bykke, L. A. Vinokyrov, M. V. Fok: Opt. i Spektroskopiya **21**, 449 (1966)
2.80   M. P. Lisitsa, Yu. S. Krasnov, V. F. Romanenko, M. B. Reifman, O. T. Sergeev: Opt. i Spektroskopiya **28**, 492 (1970)
2.81   H. Matsunami, A. Suzuki, T. Tanaka: In *Silicon Carbide–1973*, ed. by R. C. Marshall, J. W. Faust, C. E. Ryan (University of South Carolina Press, Columbia, South Carolina, 1974) p. 618
2.82   V. I. Sokolov, V. V. Makarov, E. N. Mokhov, G. A. Lomakina, Yu. A. Vodakov: Soviet Phys. Solid State **10**, 2383 (1969)
2.83   A. P. Andreev, E. E. Violin, Yu. M. Tairov, O. Ya. Fayanc: Soviet Phys. Semicond. **7**, 208 (1973)
2.84   R. M. Potter: J. Appl. Phys. **43**, 721 (1972)
2.85   A. P. Andreev, E. E. Violin, Yu. M. Tairov: Soviet Phys. Semicond. **9**, 1056 (1976)
2.86   A. Todkill, R. W. Brander: Mater. Res. Bull. **4**, S 293 (1969)
2.87   R. M. Potter: Mater. Res. Bull. **4**, S 223 (1969)
2.88   G. Borda, E. E. Violin, G. N. Violina, Yu. M. Tairov: Soviet Phys. Solid State **11**, 2051 (1970)
2.89   L. B. Griffiths, A. I. Mlavsky: J. Electrochem. Soc. **11**, 805 (1964)
2.90   V. I. Pavlichenko, I. V. Ryzhikov: Soviet Phys. Semicond. **11**, 1368 (1969)
2.91   Kh. Vakhner, Yu. M. Tairov: Soviet Phys. Solid State **11**, 1972 (1970)
2.92   Kh. Vakhner, Yu. M. Tairov: Soviet Phys. Solid State **12**, 1213 (1970)
2.93   S. Susman, R. S. Spriggs, H. S. Weber: In *Silicon Carbide – A High Temperature Semiconductor*, ed. by J. R. O'Conner, J. Smiltens (Pergamon Press, London 1960) p. 94
2.94   M. Bonke, E. Fotzer: Ber. Deutsch. Keram. Ges. **43**, 180 (1966)

2.95   E. C. Lowe: USA Pat. 3343920 (1967)
2.96   V. P. Novikov, V. I. Ionov, P. S. Spacenaya: In *Silicon Carbide* (Naukova Dumka, Kiev 1966) p. 217
2.97   E. E. Violin, A. A. Kalnin, V. V. Pasynkov, Yu. M. Tairov, D. A. Yas'kov: Mater. Res. Bull. **4**, 231 (1969)
2.98   A. A. Kal'nin, V. V. Pasynkov, Yu. M. Tairov, D. A. Yas'kov: Phys. *p-n Junctions and Semiconductor Devices* (Nauka, Leningrad 1969) p. 75
2.99   R. M. Potter, J. M. Blank, A. Addamiano: J. Appl. Phys. **40**, 2253 (1969)
2.100 R. W. Brander: Rev. Phys. Technol. **3**, 145 (1972)
2.101 A. A. Kal'nin, V. V. Lushnin, Yu. M. Tairov, I. I. Khlebnikov: Electron. Tekh. II **4**, 22 (1975)
2.102 V. M. Gusev, K. D. Denakov, M. G. Kosachenova, M. B. Reifman, V. G. Stolyarova: Soviet Phys. Semicond. **9**, 820 (1976)
2.103 A. A. Kal'nin, Yu. M. Tairov: Isvestijy LETI **61**, 26 (1966)
2.104 R. W. Brandner: Mater. Res. Bull. **4**, 187 (1969)
2.105 E. E. Violin, Yu. M. Tairov: In *Silicon Carbide–1973*, ed. by R. C. Marshall, J. W. Faust, C. E. Ryan (University of South Carolina Press, Columbia, South Carolina, 1974) p. . . .
2.106 G. F. Kholyanov, Yu. A. Vodakov: In *Silicon Carbide–1973*, ed. by R. C. Marshall, J. W. Faust, C. E. Ryan (University of South Carolina Press, Columbia, South Carolina, 1974) p. 574
2.107 R. B. Campbell, H. S. Berman: Mater. Res. Bull. **4**, 211 (1969)
2.108 A. A. Kal'nin, Yu. M. Tairov: Prib. i Tekh. Eksper. I, 143 (1972)
2.109 V. A. Barkov, A. A. Kal'nin, L. M. Kogel, Yu. M. Tairov: Prib. i Tekh. Eksper. **5**, 229 (1970)
2.110 G. V. Mikhailov, Yu. N. Nikolaev: Soviet Phys. Solid State **6**, 375 (1972)
2.111 I. V. Ryjikov, V. I. Pavlichenko, T. G. Kmita: Radiotekhnika i Electronika **3**, 6 (1967)
2.112 V. I. Pavlichenko: Soviet Phys. Solid State **8**, 984 (1966)
2.113 V. I. Pavlichenko, I. V. Ryjikov, O. R. Abdulaev: Fiz. Tekh. Poluprov. **4**, 2410 (1970)
2.114 J. M. Blank: In *Silicon Carbide–1973*, ed. by R. C. Marshall, J. W. Faust, C. E. Ryan (University of South Carolina Press, Columbia, South Carolina, 1974) p. 550
2.115 Yu. M. Altaiskii: *Property of Silicon Carbide* (Znanie, Kiev 1971) p. 3
2.116 A. Addamiano, R. M. Potter, V. Ozarow: J. Electrochem. Soc. **110**, 517 (1963)
2.117 G. A. Lomakina: Soviet Phys. **7**, 475 (1965)

3. III-V Compound Semiconductors

P. J. Dean

With 37 Figures

3.1 General

The III-V family of materials contains some of the compounds with semi-conducting properties most like those of the classical group IV elemental semiconductors germanium and silicon, fulfilling some early predictions made by *Welker* [3.1]. Our insight into and consequent ability to control these properties was initially developed through detailed study of the crystal growth, metallurgical and electronic properties of these perfectly covalently bonded elemental semiconductors. Germanium and silicon have relatively low melting points and can be readily grown in large single crystal form. Techniques such as zone melting were developed to refine these crystals to the point whereby their electrical properties could be tailored through the addition of microscopic quantities of certain key impurities which produce the well-known donor and acceptor centers. The possible existence of two dominant forms of conductivity within different regions of a given single crystal containing different major dopant species led to the design of *pn* junction rectifiers with essentially ideal electrical characteristics and the invention of the bipolar transistor. The active regions of these devices, the depletion layers between the *p* and *n* regions in which hole and electron conductivity, respectively, predominates, can be formed well within the bulk of the single crystal. They can lie well away from the external surfaces whose characteristics were difficult to assess and control in the early work. The basic principles of the energy band theory of crystals, from which the concepts of bipolar conductivity emerge, are especially appro-priate to covalently bonded semiconductors. However, key properties of relatively low electrical carrier mass, relatively long scattering times and there-fore high carrier mobilities extend also into the compound semiconductors whose electronic basis represents only a small perturbation from perfect covalent bonding. The III-V compounds come closest to this ideal. They form extended crystals in which the outer electronic shells of each III-V atom pair contribute 8 electrons to the interatomic bonding process, just as in the elemental semiconductors. The same basic crystal structure (zincblende) results, without the inversion symmetry of the elemental (diamond structure) semiconductor lattice. A relatively minor variation in the stacking sequence of atom planes along a $\langle 111 \rangle$-type axis produces an alternate form, the wurt-zite crystal structure. These differences in crystal structure, diamond→zinc-blende→wurtzite represent reductions in the crystal symmetry, reductions in the space group and consequent reductions in the degeneracy of the electronic band structure, seen most clearly at key symmetry points [3.2]. One aspect of this symmetry reduction which has been clarified only relatively recently is discussed in Section 3.3. However, the essential point here is the close approach of certain of the III-V compound semiconductors to the near ideal electronic behavior of the now classical elemental semiconductors. We must readily admit that the compound semiconductors only *approach* this ideal, in general. Their endemic drawbacks stem from the most obvious differences. It is more difficult to prepare a near-perfect single crystal of a binary semiconductor than an elemental one. There are now two atomic species to purify of impurities, which must then be controlled in their incorporation in the crystal lattice. The process becomes particularly difficult if the vapor pressure of the two atomic species is very different at the growth temperature. Special precautions are then needed to ensure near-stoichiometric growth. However, the bulk growth temperature may be inconveniently high from considerations of the constituent vapor pressure, or simply with regard to the problem of maintaining high purity in the growth environment. The compound semiconductors do have some metallurgical advantage in the wider range of possible growth techniques, bringing with them the possibility of stoichiometry control by varying routes.

With all these problems, we might well ask why there should be significant commercial interest in compound semiconductors. The answer, of course, lies in the absence of a sufficient range of useful properties within the small number of elemental semiconductors. This limitation is particularly apparent for the optoelectronic properties. We have seen in Chapter 1 that relatively efficient electroluminescence is possible only for light of quantum energy $hv \lesssim E_g$ [3.3]. This restriction shows that Ge and Si can be efficient light emitters in the near infrared spectral region at best, not in the visible. In practice they are not very efficient even for these invisible wavelengths, particularly Si which has very adverse non-radiative surface recombination, relatively long minority carrier diffusion lengths and consequently poor luminescence performance even at low temperatures. This poor performance stems partly from the fact that both Ge and Si have indirect minimum energy gaps, with low intrinsic interband absorption coefficients and consequently long radiative decay times according to detailed balance [3.3] (Chap. 1, see also Ref [3.3]). Thus, the radiative recombination route for minority carriers injected through a *pn* junction competes poorly with other, generally non-radiative, recombination processes. We shall see that these alternative recombination processes often involve crystal defects or impurities present inadvertently in practical materials and remain unrecognized to this day even in the most relevant materials. There is much interest in *visible* electroluminescence, which requires an energy gap of $\gtrsim 1.8$ eV according to our criterion. The only group IV semiconductors which meet this requirement are diamond and the single known IV-IV compound semiconductor SiC. Unfortunately, both are indirect gap and both have difficult crystal growth technologies, not well suited to the economic production of devices. No commercial devices are currently made from these two materials, although there was some adventurous early work particularly in SiC [3.3].

Turning of necessity to the III-V compounds, we encounter some promise, then a general problem which exists for other, more ionic compound materials such as the II-VI semiconductors. The promise comes in the recognition of certain III-V semiconductors whose energy gaps are large enough to support efficient luminescence well into the visible spectrum and which can also be crystallized relatively easily into n- and p-type conductivity forms to prepare electrically efficient pn junctions. Foremost among these are the binary compound semiconductor GaP, $E_g \sim 2.26$ eV at 300 K and the ternary alloy semiconductor $GaAs_{1-x}P_x$. The overwhelming majority of commercial light emitting diodes (LEDs) are currently made from these two materials, mostly from the latter. The preference for the alloy stems from the early development of a cheap crystal growth process, readily adaptable to high-yield production of crystal slices of adequate optical quality for the preparation of LEDs by Zn in-diffusion [3.4]. The crystal quality is not exemplary, as emphasized later. However, such deficiencies were masked in the initial commercial devices through the use of a compositional parameter $x \sim 0.4$, sufficiently large to produce visible (red) near gap luminescence yet small enough to permit the luminescence to retain the short lifetime characteristic of recombinations at a direct energy gap (Fig. 3.1).

Gallium arsenide is a direct gap semiconductor with excellent optical and electrical properties, but with an energy gap of only ~ 1.4 eV, much too small to support efficient visible electroluminescence. This is an aspect of the general problem mentioned above, still largely unsolved commercially and incompletely understood from an academic standpoint. Crystalline solids of large energy gaps can be obtained in two ways, both possessing severe drawbacks. We can expand away from the central column IV in a given row of the periodic table. Then we find large energy gaps only at the expense of going to relatively ionic-bonded compounds. We first notice the presence of intransigent electrical properties; poor mobilities and perhaps an inability to promote ambipolar conductivity. Then, there is a phase change to the higher coordination number of the rocksalt lattice, with generally adverse thermal and mechanical properties. There is much discussion and a certain amount of evidence, quite frequently speculative, that the problems of electronic control are related to self-compensation by the generation of electrically active lattice defects [3.5]. Other problems also exist, for example, the dual electronic role of key shallow acceptor dopants in the II-VI compounds [3.6]. The alternative solution for large energy gaps involves the selection of those compound semiconductors within a given series which possess lighter atomic constituents and, more significantly, smaller lattice constants, taken from higher rows of the periodic table. In general the minimum energy gap increases rapidly with decrease in atomic number, particularly because the direct gap at $K=0$, which lies lowest in many heavy semiconductors, is a rapidly decreasing function of lattice parameter. The adverse effect here is the very rapid increase in the growth

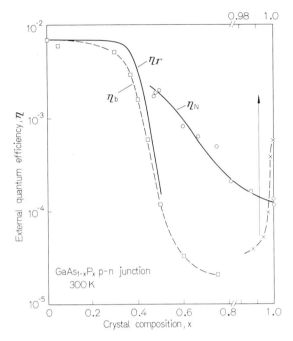

Fig. 3.1. The electroluminescence quantum efficiency η of VPE $GaAs_{1-x}P_x$ as a function of the alloy composition x. The solid curve η_Γ is the calculated contribution to η for band to band re-combinations in the N-free alloy. The calculated contribution from excitons bound to N isoelectron-ic traps has been fitted to experiment (O points) allowing for increasing "band structure enhance-ment" as the direct and indirect conduction bands approach crossover near $x = 0.4$ and for decrease in N solubility with decreasing x. The left-hand dashed curve (□ points) represents measured quantum efficiencies for N-free material [3.26]. The right-hand dashed curve (× points) was obtained from LPE $GaAs_{1-x}P_x$: N within the small range of x near $x = 1.0$ shown in the upper scale [3.29]

temperature of these semiconductors with decreases in atomic number and lattice parameter, soon reaching a level inconvenient for the economic growth of the refined crystals required for semiconductor devices. Many of the lighter atoms are very reactive and so difficult to handle on a commercial scale. For example, III-V compounds involving Al are mostly inconveniently hygroscopic. In addition, we shall see through the example of the brief discussion of the relatively wide-gap III-V compound GaN that the problem of lattice defect-associated electronic compensation occurs once again.

Thus, we must conclude that nature has been rather ill-disposed towards visible electroluminescence. The most efficient devices involve limited com-promises between wide bandgaps for wide color coverage and narrow band-gaps for ease of crystal growth in controlled amphoteric electrical form. These limitations have been discussed at length in a recent review [3.3]. In the present brief chapter, we emphasize some recent developments which give promise for the easement of some of these difficulties. A first requirement is the availability of semiconductors whose electrical properties are dominated by recognized and controlled dopants. We begin by discussing the general use of optical techniques for the recognition of impurities in semiconductors. In particular, we concentrate on those impurities which produce the shallow donors and acceptors which usually control the electrical conductivity, and yet may often be present inadvertently. About 10 years ago, it was discovered that certain key dopants, the so-called isoelectronic trap luminescence activators, can promote relatively efficient near gap luminescence even in indirect gap semiconductors, particularly those with appropriate structures in the higher-lying band extrema. We discuss some recent developments and confirmation of these ideas and the consequent exploitation of properties which now play a major role in commercial LEDs, both for GaP and $GaAs_{1-x}P_x$. Given an efficient luminescence activator, the LED will be efficient over-all only if the concentrations of centers through which competing non-radiative recombinations predominantly occur can be minimized. This aspect of the problem represents a much more difficult task for the assessment scientist. It is only very recently, being tackled systematically, through the application of newly developed diagnostic techniques. Some of these techniques are briefly described, and the significance of their early results is discussed. Much of the current evidence for the role of extended defects in quenching luminescence performance comes from recent widescale attempts to improve the short-term degradation characteristics of GaAs – (GaAl)As heterostructure injection lasers. This work is also briefly described, although injection lasers are generally outside the scope of this chapter. We show that quantum effects of carrier confinement can be seen in some recent device structures, particularly in electrical properties of insulated gate field effect transistors and in optical properties of double heterostructure lasers with sufficiently narrow active regions. These effects are of interest in their own right and because of the information they provide on key electronic properties of the device structures. We continue with a brief description of GaN, the most interesting "exotic" III-V compound semiconductor from the electroluminescence standpoint. In conclusion some general comments are made about the prospects for the future of electroluminescence in III-V semiconductors.

3.2 Optical Techniques for Impurity Recognition

The current very comprehensive understanding of the electronic properties of shallow, electrically active impurities in the important LED-prototype III-V compound semiconductor GaP was largely obtained through the study of appropriately refined and judiciously back-doped crystals using optical spectroscopy. The binding energies of the more important of these, together with some of the isoelectronic trap luminescence activators discussed in Section 3.3, are shown in Fig. 3.2. It was natural to consider the use of optical

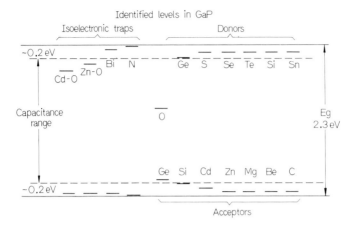

Fig. 3.2. A schematic summary of the energies of many of the impurities determined from low temperature photoluminescence spectra of GaP. This technique is most suited to levels close to the band edges. In general, deep traps with energies closer to the middle of the forbidden gap are best determined by methods which do not rely on significant luminescence in the capture or recombination processes. This is particularly true for centers in which a change of charge state is accompanied by a large local lattice relaxation, probably mainly of associate type. The various capacitance techniques described in the text are particularly applicable to the study of such deep traps [3.178]

spectroscopy for the analysis of the wider gap semiconductors intended for optoelectronic devices, just as the earlier-developed narrower-gap semiconductors were analyzed by electrical transport. The optical techniques have many advantages. On the practical side, the specimen preparations are minimal. No contacts are required if the luminescence is excited by an optical or electron beam. Data analysis is fast and straightforward. The parameter which distinguishes different impurities, the depth of the traps they induce for electronic particles, can appear directly in optical spectra taken at a single fixed sample temperature. In contrast, by transport measurements, such trap depths are derived from a relatively tedious analysis of the temperature dependence of electrical data taken over an appropriately wide range. Even then, the electrical data provide information on trap depths only for the majority type of species present in a given sample, donor or acceptor. The characteristic signatures of donors and acceptors are jointly present in a single typical optical spectrum, for example the d−a pair spectra shown in Fig. 3.3. The energies of the many lines shown in Fig. 3.3 are directly related to the energy sum $E_A + E_D$ through the simple relationship [3.7]

$$hv = E_g - (E_A + E_D) + e^2/_{\varepsilon R} + f(R).$$ (3.1)

The correction term $f(R)$ is significant only for relatively close pairs. The presence of more than a single donor or acceptor species is sensitively detected through the appearance of additional series of sharp no-phonon lines in these

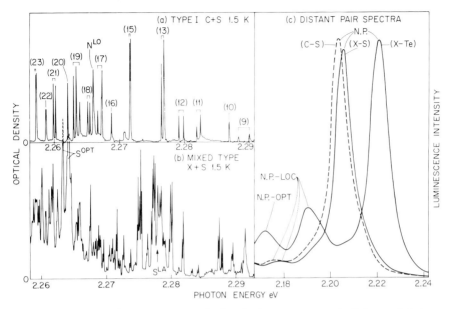

Fig. 3.3a–c. Portions of donor-acceptor pair photoluminescence spectra in GaP recorded photo-graphically (a) from GaP:S,C and (b) from GaP:S,X. The energy scale of (a) is shifted 2 meV to the right to superpose transitions at the given pair separations. The numbers in (a) are an index of these separations. The spectra in (c) are recorded photoelectrically with very low excitation intensities to show the displacements of the distant pair peaks for the indicated impurities. X is an unidentified (axial) acceptor [3.12]

'edge luminescence' spectra [3.8]. This is possible even when the trap depth is abnormally large, as for the donor O_P in GaP. The many donor and acceptor ionization energies shown in Fig. 3.2 were derived from these energy sums for d–a pair spectra, given independent knowledge of one value of E_D or E_A. This knowledge was obtained from infrared photoionization spectra of donors [3.9] and approximately confirmed through electrical transport analysis of the acceptor Zn in carefully prepared p-type GaP [3.10]. The absolute accuracy of this knowledge is limited by uncertainties in the analysis of the donor states, due to complexities in the conduction band structure which have only recently been clearly identified [3.11]. These complexities effectively make the indirect conduction band minima much more anisotropic than they would be with minima exactly at X as in Fig. 3.8. Most recent analyses [3.11a, 11b] suggest that the set of donor energies in Table of Ref. [3.3] are underestimated by about 3 meV. In addition, structured photoionization spectra of acceptors recently available for the first time in GaP [3.11c] suggest that the set of E_A are low by ~ 8 meV. Thus, from (3.1) E_g should be increased by ~ 11 meV at 0 K.[1] This approach gives the best available estimate of E_g since only the ex-citon gap is accurately measured from the indirect optical absorption edge

_____
[1] See *Notes Added in Proof*, however, p. 126, Note (1).

[3.11 d]. However, the energy differences between different donors and acceptors are established to better than ± 0.1 meV directly from the d−a pair spectra, adequate for most purposes including positive impurity recognition. An advantage of the d−a pair spectra here is that the no-photon transition energies depend only upon the impurity *ground state* ionization energies which, apart from the easily computed $e^2/\varepsilon R$ factor in (3.1), are far less perturbed by interimpurity interactions than the activation energies measured in electrical transport.

The presence of different donor (or acceptor) species can also be recognized from appropriate shifts in the broad overall spectral peak due to electron-hole recombinations on very distant, unresolvable d−a pairs, also shown in Fig. 3.3. These peaks are relatively narrow if the spectra are recorded either at very low excitation intensities or long time delays after pulse excitation, when the higher energy luminescence at relatively close pairs becomes de-emphasized. For d−a pairs with very small energy sums $E_A + E_D$, the relatively close d−a pairs can no longer bind either electronic particle and only the unresolved transitions at distant pairs appear in the spectra. Typical shallow donors and acceptors in the relatively narrow gap II-VI compound semiconductor CdTe have $E_A + E_D$ just critically large enough for transitions at discrete d−a pairs to be resolved, rather poorly [3.12].

Further decrease of E_g brings a decrease in m_e^* and E_D giving low $E_A + E_D$, large donor Bohr radii and consequent problems in the optical saturation of very distant d−a pairs in direct gap semiconductors. Thus, discrete d−a pairs

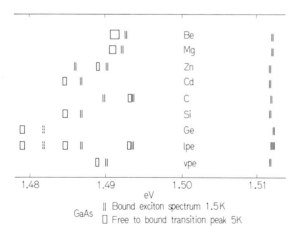

Fig. 3.4. A schematic representation of luminescence lines due to the radiative recombination of excitons bound to the indicated neutral acceptors (right) and satellites due to transitions in which the acceptor is left in the lowest S-like excited state, "two-hole" transitions (left). The boxes to the left represent bands due to the radiative recombination of free electrons at the same neutral acceptors. The lowest two entries show typical spectra for refined, undoped LPE and VPE material [3.13]

have never been resolved for the narrower gap semiconductors GaAs and InP, although the distant d−a pair peaks are readily distinguished in the edge luminescence of refined crystals at low temperatures. Great increase in the optical pumping rate promotes delocalization of electrons at the very shallow donors and the emergence of free (electron) to bound (hole), f−b luminescence bands before the discrete d−a pair lines can be seen in the same energy region. These distant d−a pair and f−b bands have been recently used to distinguish and identify many shallow acceptors in GaAs [3.13] (Fig. 3.4) and a few in InP [3.14]. Figure 3.4 also shows that these identifications have been made far more precise by the use of certain "two-hole" bound exciton satellites involving transitions to excited states of the neutral acceptors at which the excitons are bound. The complete range of normal shallow acceptors in GaAs has been identified in this way. Detailed spectroscopic information also exists on a few deep acceptors such as Sn [3.15], Mn [3.16] and Cu [3.17]. The no-phonon bound excitons can also be used for chemical analysis in Si and GaP, where the exciton localization energy E_{Bx} is a monotonically increasing function of the ionization energy of the donor or acceptor, the Haynes' rule. It has been recently shown that this behavior does not apply in GaAs [3.13] and InP [3.14], where E_{Bx} is nearly independent of E_A (Fig. 3.4). Antibinding effects of d state contributions to the hole-hole correlation energy may prevent the exciton binding which is normally possible for neutral donors or acceptors [3.14a].

Spectroscopic study of impurity effects through such band edge luminescence is experimentally very convenient. However, it is not possible to use the technique for identification of the shallow donors in semiconductors with low m_e^* like GaAs and InP where the entire range of donor ground state chemical shift is ∼0.1 meV and the consequent spectral displacements are hard to unravel from the wealth of fine structure associated with shallow excited states of the donor bound exciton. The $1s − np$ transitions seen in the photoionization spectra of these donors are free from such complications. The chemical shifts are still small compared with the spectral linewidths, however (Fig. 3.5). The spectroscopy is much more difficult in the relevant, far-infrared region. The usual technique is to apply the photothermal process to enhance the photoconductivity contributions from excitation peaks to high donor excited states [3.18]. The spectral lines narrow in a high magnetic field and the scale of the central cell shifts is enlarged due to compression of the bound electron wave function (Fig. 3.5). Despite much study, summarized in Fig. 3.5, some of the identifications of major contaminant species still remain tentative. This photoconductivity method is complicated by the necessity for good, low noise electrical contacts but has the advantage that the signals are proportional to the relative concentrations of different donors, which is not always true in luminescence. Particular complications occur in GaP, due to the large difference in oscillator strength f of optical transitions associated with donors on the Ga and P lattice sites caused by the form of the Bloch part of the electron wave function [3.19]. In luminescence, the relative strength of d−a and bound exciton transitions involving the low f Si_{Ga} donor and high f S_P donor in GaP

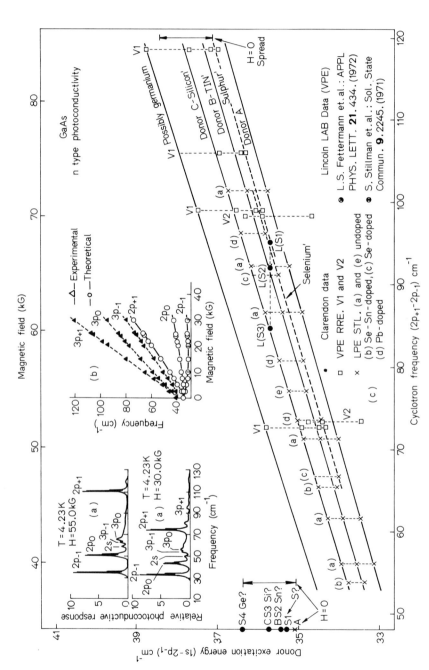

Fig. 3.5. The main figure shows the variation of the 1S-2P transition energy with the magnetic field and cyclotron frequency eH/m_e^* for a variety of epitaxial GaAs, refined and sometimes lightly back-doped as indicated. The splittings due to chemical shifts between different donor species, very small in GaAs, become slightly larger and better defined at high fields (comparison with H = O spread at extreme right). The chemical identifications are still tentative [3.181]. The insets show the behavior of the complete donor photoexcitation spectrum [3.179]. The transitions to the 3p states are incorrectly identified [3.180]

depend not only on $(N_d)_{Si}/(N_d)_S$ but on the balance of effects due to the optical pumping rate and the total as well as relative impurity concentrations. The discrete d − a pairs are tending to optical saturation in Fig. 3.6a because of low carbon concentration [C], while the Si − C pairs (dashed lines) are weak compared with S − C (full lines) because of the large ratio of oscillator strengths despite a near unity ratio of the total intensity of the Si and S donor bound excitons. The pair lines become desaturated with increase in [C] (Fig. 3.6b) and the Si − C pairs are then seen more clearly despite a lower ratio of [Si]/[S] as judged from the donor exciton spectra. The relative intensities depend upon ratios of exciton and carrier capture cross sections in these unsaturated spectra. Further increase in [Si] causes the S − C pairs to weaken considerably, while

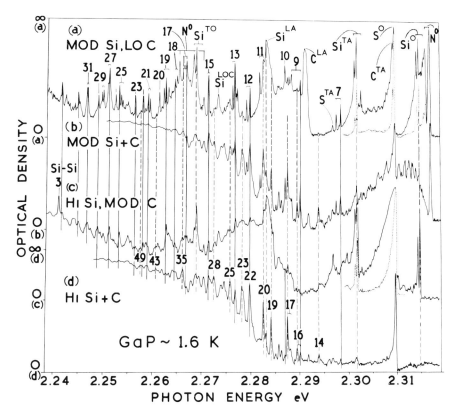

Fig. 3.6. Portions of the edge photoluminescence spectra of GaP doped with the shallow donor Si and the shallow acceptor C as indicated, recorded photographically. The donor S is present inadvertently. The vertical solid lines denote S-C donor-acceptor pair transitions, dashed lines indicate Si-C pairs and the numbers are an index of donor-acceptor separation. Optical saturation effects and the large difference in oscillator strengths of no-phonon transitions involving the S_P and Si_{Ga} donors produce complications in the assessment of the numerically dominant shallow donor from these spectra, although the effects of these two donors are clearly resolved [3.34]

the Si donor spectrum becomes visible (Fig. 3.6c). Finally, the bound exciton transitions become very weak because of tunneling to the more localized d−a pair states when both [Si] and [C] are large (Fig. 3.6d).

The $1s$-np donor excitation spectra are not subject to these problems of widely different f values, but are difficult to study in GaP because of obscuration by multiphonon absorption [3.9] and are also difficult to interpret because the camels' back effect [3.11] described in Section 3.3. Another technique, convenient in GaP because of an appropriate slight excess of the donor interbound state excitation energies ΔE_D over the zone center longitudinal optical $LO(\Gamma)$ phonon energy, makes use of the slight difference in energy of an LO phonon in the vicinity of neutral donor. If $\Delta E_D > \hbar W_{LO}$, the phonon energy is slightly reduced by the dielectric screening produced by the bound electron. The energy-shifted phonon may be seen in both Raman scattering [3.20] and Reststrahl reflectance spectra [3.21]. In either case, the strength of the impurity mode increases with donor concentration with essentially the same coefficient of proportionality for Ga and P site donors (Fig. 3.7). This coefficient can be determined numerically from a simple theoretical treatment [3.20] knowing the Fröhlich interaction constant for the GaP lattice and the donor excitation energies. Impurities and their associates can also be detected by infrared spectroscopy using the special vibrational modes which are caused by the mass defect with respect to the host atom they replace and the local changes in force constant [3.21a]. This method has several advantages compared with luminescence,

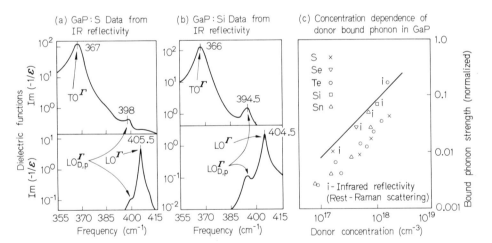

Fig. 3.7a−c. The dielectric response functions of GaP doped with the donors S [spectra (a)] and Si [spectra (b)] showing the additional mode of symmetry identical to the normal LO^Γ mode which appears at slightly lower energy depending upon the donor photoexcitation energies. Part (c) shows that the oscillator strength of this transition, observed in Raman scattering as well as infrared reflectivity, is essentially as large for the Ga-site donors Si and Sn as for the P-site donors S, Se, and Te, unlike the relevant edge photoluminescence transitions [3.21]

particularly that the spectral response is directly proportional to the impurity concentration. Disadvantages include the need for calibration, unlike the bound phonon technique just mentioned, and the fact that the method is sensitive only for impurities with advantageous mass defects and then only for relatively high concentrations $\gg 1$ ppma. In addition, measurements for electrically active species are possible only after the excess carriers have been removed either by in-diffusion of suitable compensating impurities or by the introduction of deep traps associated with radiation damage. These treatments mean that the method is destructive and consequent changes of state of the impurities may complicate the interpretation of the infrared spectra. The method is well suited to the study of high concentrations of impurities, where associative effects are most pronounced and are often best detected by this method.

3.3 Band Structure of GaP and the Isoelectronic Trap Luminescence Activator

Many of the basic band structure parameters of GaP and GaAs have been derived from optical spectroscopy, in particular the general features adjacent to the forbidden gap sketched in Fig. 3.8. As noted in Section 3.2, GaP is an indirect gap semiconductor with a set of equivalent conduction band minima near X and a degenerate valence band maximum at the center of the Brillouin zone, symmetry point Γ. Recent analysis of a satellite series prominent in the luminescence spectra of excitons bound to shallow acceptors in GaP has shown that it arises from exciton decay with simultaneous scattering of the weakly bound electron between conduction band valleys on the same $\langle 100 \rangle$ symmetry axis [3.11] g-scattering (Fig. 3.9a). The energy of the LA phonon emitted in this process provides an accurate estimate of the displacement wave vector K_0, $0.047 \, K_{\langle 100 \rangle}^{max}$. Thus, GaP has the complicated "camels' back" form of conduction band structure shown in Fig. 3.9b, with $\Delta E \sim 10$ meV. This value is comparable with both the kinetic energies for electrons in typical electrical transport and cyclotron resonance studies and the ionization energies of the excited states of donors. Its existence vastly complicates the analysis of these phenomena compared with Si where $K_0 = 0.15 \, K_{\langle 100 \rangle}^{max}$ and electrons within the three pairs of adjoining valleys are well decoupled, $\Delta E \sim 0.47$ eV.

Optical spectra have also been used to determine the properties of a series of luminescence activators in GaP, particularly isoelectronic trap centers such as Zn_{Ga}-O_P and N_P which are used, respectively, in red and green (or yellow) LEDs [3.3]. Both of these isoelectronic traps have very electronegative cores and interact primarily with electrons to form the bound excitons which recombine with efficient luminescence. These states contain a single electron and hole. Thus there is no possibility of an "internal" Auger effect, such as is known to severely quench donor and acceptor bound exciton luminescence in indirect gap semiconductors like GaP [3.22]. We shall concentrate here on N_P. There has been some very recent work on the basic optical properties of this luminescence activator and it has become very widely used in commercial LEDs, $GaAs_{1-x}P_x$ as well as GaP (Table 3.1). The strong but short range attractive potential of the N_P core for electrons induces a very shallow bound state ~ 10 meV. There has been some debate about the existence of a bound state for an electron alone. The exciton bound state is clearly established from the optical spectra. The most recent theoretical calculations support the idea that the electron is just bound at N_P in GaP and more strongly bound in $In_xGa_{1-x}P$ [3.23]. Recent experimental evidence from the excitation of N exciton absorption by strong optical pumping above the bandgap of GaP gives evidence for single electron as well as exciton capture [3.24]. Whether the single electron is bound or not, its wavefunction in the bound exciton state has a major contribution near $K=0$, beneath the relatively low-lying subsidiary conduction band minimum there (Fig. 3.8). The $\Gamma - X$ valence band energy interval is much larger, so there is no correspondingly large spread in hole

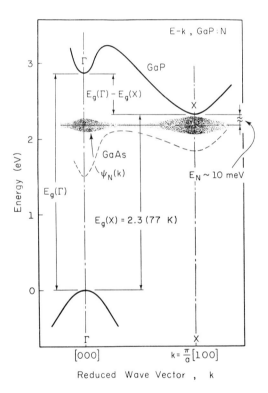

Fig. 3.8. The lowest conduction bands and highest valence bands of GaP (solid) and GaAs (dashed) in the Γ-X direction of the reduced zone. The wide spread of the wave function of an electron bound to the indicated shallow isoelectronic trap N_P is indicated schematically for GaP. The abnormally large proportion near $K=0$ for this type of trap produces the characteristically high oscillator strength for radiative recombinations with weakly bound holes [3.46]

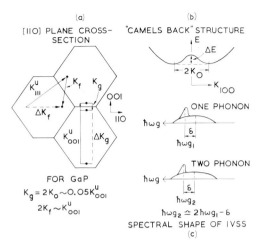

Fig. 3.9. (a) [110]-plane cross section of several adjacent Brillouin zones in a semiconductor like GaP with conduction band minima just inside the zone boundary at the X symmetry points. The X points lie at the center of the short sides of the distorted hexagons. Wave vectors K_{111}^u and K_{001}^u are principal vectors of the reciprocal lattice and represent "Umklapp" scattering processes, while K_f and K_g are the wave vectors required for f and g-type intervalley scattering. The quoted values of K_f and K_g are derived for GaP from analysis of the intervalley scattering satellites in the luminescence spectra of excitons bound weakly at shallow acceptors, shown schematically in (c) in fundamental and first overtone. Part (b) shows a [100] orientation section of the "camel's back" structure resulting from the two minima positioned symmetrically about X in the Γ-X direction [3.11]

Table 3.1. Current Status of Commercially Available Visible LEDs

LEDs	Color	Peak emission wavelength (Å)	Luminous output (lumen/watt)	External quantum efficiency		Best luminous efficiency (lumens/watt)
				Best (%)	Commercial (%)	
GaP:Zn,O	Red	6990	20	15	2.0 −4.0	3.0
GaP:N	Green	5700	610	0.7	∼0.05−0.1	4.2
GaP:NN	Yellow	5900	∼450	0.1	∼0.05	0.45
GaAs$_{0.6}$P$_{0.4}$	Red	6490	75	0.5	0.2	0.38
GaAs$_{0.35}$P$_{0.65}$:N	Orange	6320	190	0.5	∼0.2	0.95
GaAs$_{0.15}$P$_{0.85}$:N	Yellow	5890	450	∼0.2	0.05	0.90

[3.27]

wave function for a hole-binding isoelectronic trap such as Bi_P. The experimental ratio f_N/f_{Bi} is in good accord with simple theoretical prediction $\{(E_\Gamma)_c - (E_x)_c/[(E_\Gamma)_v - (E_x)_v]\}^2$, about 25 according to the known band structure of GaP. Since the hole in the bound exciton is weakly bound to the electron at N_P by long range forces, the oscillator strength of the indirect transition is mainly determined by the value of $|\psi(k)|^2_{k=0}$ for the bound electron. This quantity is more than 100 times greater for binding at the 10 meV isoelectronic trap N_P compared with a 100 meV donor such as S_P [3.25]. Recent calculations [3.26] have shown that there is a further $\sim$1000-fold increase for $GaAs_{1-x}P_x$ for $x=0.45$ (Fig. 3.10), just above crossover from direct (GaAs-like) to indirect (GaP-like, Fig. 3.8) behavior. The very large value of this "band structure enhancement" of the N oscillator strength [3.25, 26] gives a corresponding decrease of radiative lifetime τ_r to a value comparable with intrinsic direct transitions, $\sim$1 ns. The non-radiative lifetime τ_{nr} is still much shorter than this due to the shunt path recombinations discussed later. However, this large decrease of τ_r has rendered obsolete the assumption usually made in the analysis of the compositional dependence of unactivated $GaAs_{1-x}P_x$, namely that $\tau_{nr}/\tau_r=0$ for electrons which recombine from states associated with the indirect (X) minima (Fig. 3.1). Analysis with a simple kinetic model for the N-activated alloy yields the theoretical curve labeled η_N in Fig. 3.1, in fair agreement with experiment. This analysis, like the data in Fig. 3.10, takes account of the increase of exciton localization energy with decreasing x deter-

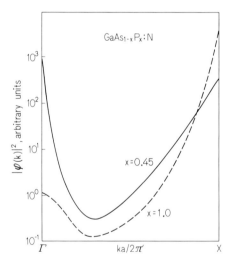

Fig. 3.10. The modulus of the wave function $|\phi(k)|^2$ of an electron bound to a shallow N isoelectronic trap in GaP and $GaAs_{0.55}P_{0.45}$ along the Δ symmetry line in the reduced zone. Note the very rapid increase near $K=O$ with decrease in [P] in the alloy above even the abnormally large value characteristic of N in GaP (Fig. 3.8). This results in a larger value at Γ than at X for the alloy shown, close to conduction band crossover [3.26]

mined from the optical spectra. Also, the fit of η_N is improved if [N] at $x=0.5$ is set to 20% of the value for $x=1.0$, i.e., there is a significant decrease of N solubility in the alloy.

The very large increase of η_N compared with η_b gives useful output for GaAs$_{1-x}$P$_x$:N LEDs over the whole color range red→yellow (Fig. 3.11) and is currently under commercial exploitation (Table 3.1). *Campbell* et al. [3.26] emphasize the still greater potential for In$_{1-x}$Ga$_x$P:N in the yellow (Fig. 3.11). However, there are great problems of realizing this potential with a commercially acceptable LED technology in this metallurgically difficult alloy. Two factors are worthy of further emphasis for GaAs$_{1-x}$P$_x$. First, the value of τ_{nr} for indirect alloy is very short in typical commercial material, indicating excess quantities of the deep states responsible for the shunt path compared with the binary compound, as is consistent with direct evidence from preliminary deep level studies [3.27]. There is some recent evidence that this may not be an inherent problem of the ternary alloy, at least for VPE material [3.28]. However, the precipitous decrease of η with x noted near $x=1.0$ in Fig. 3.1 suggests the presence of severe shunt path recombination problems in some types of

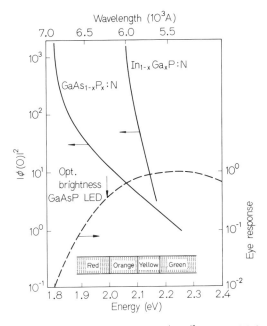

Fig. 3.11. The probability density $|\phi(0)|^2$ at $K=O(\Gamma)$ for an electron bound to N in GaAs$_{1-x}$P$_x$ and In$_{1-x}$Ga$_x$P, shown as a function of energy which varies with change in alloy compositional parameter x. The spectral variation in eye response for a normal observer is shown dashed, while the color spectral scale is added for convenience. The variation of $|\phi(0)|^2$ is particularly rapid for In$_{1-x}$Ga$_x$P and lies further in the orange-yellow, but this alloy has difficult metallurgical problems. The optimum brightness for a GaAs$_{1-x}$P$_x$: N LED lies near the red-orange boundary, but the variation is relatively small throughout the red-yellow spectral range [3.26]

alloy material, here LPE [3.29]. Some have found evidence for a similar dramatic effect in the VPE alloy [3.30], but it is not understood at present. Anyway, it is clear that if τ_{nr} of the indirect alloy could be improved to near the value for good binary GaP, say ~ 50 ns, then the extremely large value of the band structure enhancement for the N activator in the indirect alloy would produce very worthwhile further gains in LED efficiency in the red, orange and even yellow.

3.4 New Effects of Nitrogen in GaP and GaAs$_{1-x}$P$_x$

The model of binding at isoelectronic traps described in the previous section has had strong support from a recent study of optical excitation spectra for the luminescence of excitons bound relatively tightly to a number of NN_m associates [3.31]. The hole is weakly bound in the long range Coulomb field of the electron bound to the neutral core. Such a system has been described as an isoelectronic acceptor [3.32]. If this is so, the hole should exhibit a series of acceptor-like excited states. The usefulness of this description is proved in Fig. 3.12, where each series of principal excited states observed for associates with $m < 6$ is well described by the expression $E_n = E_1 n^{-1.8}$, where E_n is the hole binding energy in the n-th state. This quasi-Rydberg has been predicted from effective mass theory for acceptors in zincblende semiconductors. The ~ 9 meV energy shift between theory and experiment in Fig. 3.12 arises from an overestimate for the ground state binding energy of an isoelectronic acceptor E_1 by a theory which assumes a fully localized negative core charge. This assumption becomes still poorer with increasing m. The entire basis for this description finally breaks down when the experimental electron plus hole binding energy $E_e + E_h$ becomes smaller than the theoretical value for E_h alone. An excitonic model must then be used. Correlation is very important in the motion of the electronic particles in these weakly bound excitons as shown, for example, by the large binding energy of the second exciton in the N_P bound excitonic molecule in GaP [3.33]. Rather direct experimental confirmation [3.34] of strongly correlated electron-hole motion in N_P and other shallow bound excitons where $m_e \sim m_h$ has been obtained recently from measurements of their diamagnetic energy shifts (Fig. 3.13). These shifts are anomalously small, about 10% of the prediction on a simple hydrogenic model assuming well separated core and orbital electronic charges for N_P. This quenching is attributed to a "neutral current" effect of strong electron-hole correlation in the electronic motion [3.35]. The quenching effect is so large that when a large exciton binding energy is mainly due to *one* electronic particle, as is frequently so, the diamagnetism may be larger than for excitons with much smaller $E_e + E_h$!

There are many cooperative effects of impurities in optical spectra of semiconductors. Some are well explored, for example the donor-acceptor pair spectra discussed in Section 3.2. Even here, there have been recent new developments. *Hutson* [3.36] has demonstrated the inadequacy of existing theories of the fine structure exhibited within each scalar d−a pair separation. Theories which emphasize effects in the luminescence transition final state have failed to reproduce the magnitudes of the splittings with realistic material parameters, for example delocalized ionic charges or piezoelectric constants. The splittings apparently arise from *initial state* interactions, in which the difference in the donor and acceptor wave functions produces splittings in the interaction

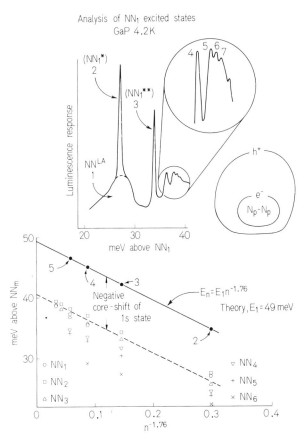

Fig. 3.12. The upper part shows a portion of the photoexcitation spectrum of photoluminescence at the NN_1 isoelectronic trap in GaP, dominated by the indicated transitions to a series of electronic excited states. The lower part shows that the excitation energies for several of the deepest NN traps are well described by quasi-Rydberg series whose dependence on the quantum number n of the excited state lies close to theoretical expectation. The theory is based upon a description of the bound exciton as an "isoelectronic acceptor" with a relatively weakly bound hole, as indicated schematically at the right center. The large uniform shift between experiment and theory, which is nearly identical for NN_1-NN_3, is attributed to a reduction in hole binding energy due to the finite extent of the wave function of the bound electron, mainly noticeable for the ground state of the "isoelectronic acceptor" [3.31]

energies of the bound particle with the multipole sources of the opposite polarity. The multipole potential arises physically from the dielectric polarization of the four nearest neighbors of each substitutional impurity ion.

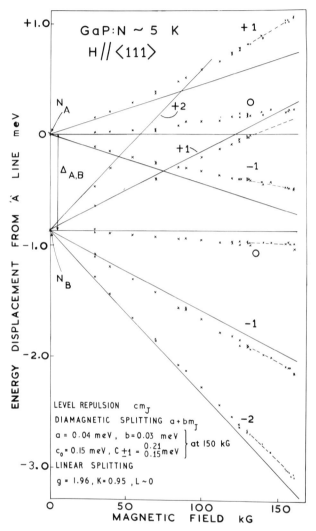

Fig. 3.13. The magnetic splittings of the transitions due to radiative recombination of excitons bound to the N isoelectronic trap in GaP. The two transitions at zero field N_A, N_B result from J-J coupling of the bound electron and hole to produce $J=1$ and $J=2$, with a $J=0$ transition final state. The linear Zeeman effect observed at low fields is discussed in [3.33]. These data show the small nonlinear effects significant above ~ 50 kG. There are contributions from level repulsion and diamagnetism, both dependent upon the magnetic quantum number m_J as indicated. There is no significant repulsion for the $m_J = \pm 2$ initial states of N_B because the magnetic field does not mix these states with those from N_A to first order (electric dipole transitions) [3.34]

Some other types of interimpurity spectra remain rather less well understood, despite recent studies. Such are the "undulation spectra" characteristic of GaP doped with N and neutral acceptors (Fig. 3.14). The original interpretation in terms of an interference effect in the wave function of the electron bound to N and perturbed by a nearby neutral acceptor [3.37] reproduces the experimental undulation peak energies quite well but seriously underestimates the observed oscillation amplitudes. *Street* and *Wiesner* [3.38] have recently criticized two subsequent models and produce evidence to support the original view that each undulation peak contains luminescence from a number of scalar d−a separations. However, they hold that the spectral structure arises mainly from vagaries in the envelope of total pair state number density as a function of the N−a separation−a probably just but perhaps rather mundane interpretation! It should be noted here that the structure associated with the

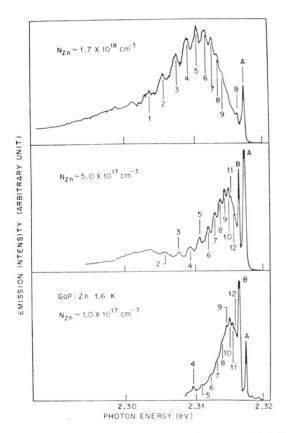

Fig. 3.14. Portions of the edge photoluminescence of GaP:N,Zn showing the A, B bound exciton lines due to isolated N_P isoelectronic traps and an associated broad low energy wing containing undulatory structure. This wing becomes dominant at high [Zn] and is due to bound exciton recombinations at N perturbed by nearby neutral Zn acceptors [3.37]

donor bound excitons, similar in cursory examination [3.39] is appreciably more complex in its spectral dependence on impurity incidence. The exact spectral form seems to depend on the nature of *all* donors and acceptors present in significant concentration. In particular it does not depend only on the P site donor which produces the dominant bound exciton no-phonon luminescence, but also on Ga site donors which also may be present [3.34]. The centers which perturb this donor exciton cannot be isolated ionized donors or acceptors because the displacement energies are too small. Probably they are d−a associates and possibly the main perturbation is due to nearest neighbor associates which may form in considerable excess due to the strong Coulomb inter-ion attraction, like the well-known associates involving the deep donor O_P which are efficient activators for red luminescence in GaP [3.3]. It is difficult to be certain about the validity of this speculative assignment. The effect is significant if correct, since these spectra then provide by far the easiest way of detecting such preferential association when the $E_A + E_D$ sum is too small to bind an exciton at the nearest neighbor so that these associates cannot luminesce in their own right and the impurity masses or concentrations are inappropriate for the use of the local mode technique [3.21a].

There have been a number of recent observations of basic properties of the important luminescence activator N in $GaAs_{1-x}P_x$. It has long been known that the N and NN_m trap depths increase with decreasing x [3.40]. This behavior has now been determined over the whole compositional range up to x_{N_c} where these impurity energies cross the Γ_1 conduction band minimum (Fig. 3.15)[2]. Because of the appreciable (nonlinear) increase of E_N with decrease in x, the composition for crossover of the N level and the Γ_1 conduction band minimum x_{N_c} lies significantly below the composition x_c where X and Γ_1 conduction band minima cross. *Campbell* et al. [3.41] have observed a considerable downshift in photoluminescence peak energies on addition to Zn to $GaAs_{1-x}P_x$:N. The shift is nearly constant at ~ 30 meV in the direct gap region (Fig. 3.15) and is identified with $(E_A)_{Zn}$ due to electron f−b recombinations at neutral Zn acceptors. However, the energies are also lowered in the region just *above* x_{N_c}. This decrease may arise from the predominance over normal N bound exciton luminescence of the recombination of electrons bound to N with holes bound to nearby Zn. The oscillator strength ratio

$$f_{Zn,N}/f_N \propto (a_{ex}/a_{Zn})^3 \exp(-2R/a_{Zn}) \tag{3.2}$$

where a_{ex} and a_{Zn} are the radii of the holes at the N exciton and Zn acceptor and R is the N−Zn separation. Since $(E_A)_{Zn}$ increases by $\sim 2X$ in going from direct to indirect gap, then the corresponding decrease in a_{Zn} accounts for a switch back to luminescence independent of Zn as x increases well above x_c (Fig. 3.15).

A most important point, recently clarified, concerns a "resonant enhancement" in interband oscillator strength alleged to occur when E_N lies just above

---

[2] See *Notes Added in Proof*, p. 126, Note (2).

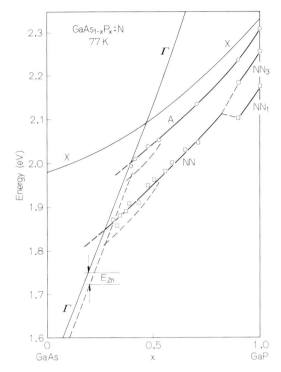

Fig. 3.15. The variation of transition energies associated with the N isoelectronic trap with the crystal composition in $GaAs_{1-x}P_x$ with (solid lines) and without (dashed) the shallow Zn acceptor. Transitions involving holes bound to the Zn acceptor are favored even in the indirect gap range just above the Γ-N crossover near $x = 0.4$ [3.41]. See *Notes Added in Proof*, p. 126, Note (2)

E_Γ. Some calculations [3.42] implied that N is an efficient luminescence activator to the point of enhancing laser action even well into the *direct* gap compositional range because of this resonance effect. However, recent studies of the effect of N addition show that such N-doped material has a significantly higher threshold for laser action [3.43]. The mode spacing in lasing material [3.44] can be used to derive the energy variation of effective refractive index with an accuracy sufficient to show the perturbing effect of the N (Fig. 3.16). The asymmetric line shape arises from the expected Fano interference mechanism between the N resonance and the underlying intrinsic absorption continuum present even though the N produces a local peak in the density of states. According to a recent re-examination of the theory [3.42], the N level is expected to produce a local quenching of luminescence, just as observed in Fig. 3.16. This reconsidered view brings us back to an earlier "hand-waving" idea that the excessive spread of wave function produced by binding in a short range potential can only be deleterious to electron-hole recombination in a *direct* gap semiconductor, as emphasized by the relatively slow decay observed for neutral luminescence activators compared with those having weak,

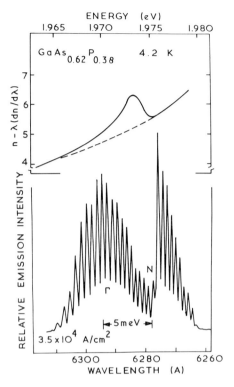

Fig. 3.16. Luminescence intensity and effective index of refraction determined as functions of energy from the longitudinal mode spacing in the N-doped $In_{1-x}Ga_xP_{1-z}As_z/GaAs_{1-y}P_y$ hetero-junction laser diode with $x \sim 0.7$, $y \sim 0.38$, $z \sim 0.01$. The N trap is ~ 5 meV above the Γ conduction band minimum for this value of y, near the steep edge of the local increase in refractive index dispersion associated with the presence of N traps. The refractive index in N-free material follows the dashed line. The large decrease in luminescence intensity near 1.975 eV shows that the resonant N level inhibits band to band recombination by broadening the wave-vector distribution of the state from which the electron recombines [3.44]

Coulombic binding [3.45]. This is a natural consequence of the opposite, beneficial role played by isoelectronic traps in *indirect* gap semiconductors. There, the band structure enhancement effect described in Section 3.3 still exists and becomes resonantly large as x_{N_c} is approached. This resonance allows stimulated emission well on the indirect side of x_c, for example to at least $x = 0.47$ at 77 K in externally pumped $GaAs_{1-x}P_x$:N [3.46].

Efficient near band gap luminescence in $GaAs_{1-x}P_x$ has recently been achieved with N ion implantation [3.46a]. The intensity of 77 K luminescence of indirect $GaAs_{0.48}P_{0.52}$ can be enhanced to a value comparable with direct gap material, perhaps $\sim 10^3$ times larger than comparable unimplanted material, like the results for N activation in growth doped material (Fig. 3.1), if implantation is carried out at 350 °C with subsequent annealing of radiation damage at 800 °C. This ion-implantation technique has appreciable diagnostic interest for the behavior of N in III-V semiconductors. However, it is unlikely to be of direct value for LED preparation. The plateau of the implanted N extends only ~0.5 μm below the surface even with 220 keV ions, much too shallow for the fabrication of efficient LEDs due to the influence of surface recombination [3.3]. No evidence has been obtained for significant post-implantation diffusion of N. Radiation damage is greatly reduced in LPE GaP if N-implantation is performed at 500 °C [3.46 b] and residual damage is largely annealed at 700–800 °C as indicated by He backscatter and the restoration of shallow d – a pair luminescence. However, the N is then in some paramagnetic form, perhaps interstitial and possibly related to the behavior of N in LEC GaP [3.97]. The N is substantially converted to P lattice sites only by anneal at 900–1000 °C. Then, the spectral form and intensity are comparable with growth doping, although the former still show greater influence of strain in the implanted samples.

3.5 Recombination Kinetics and Non-radiative Pathways

The relatively poor electroluminescence efficiencies η_{EL} listed in Table 3.1, particularly of the commercial devices, are governed by low bulk luminescence efficiencies. In a good LED, the majority of the luminescence results from radiative recombination of minority carriers injected into bulk material within a few diffusion lengths of the depletion layer, luminescence efficiency η_R. Often, the light is designed to appear predominantly from one side of the pn junction. Usually, the relevant electrical injection efficiency η_E can be arranged to be at least ~50% [3.47]. The efficiency η_{EL} is given by the product

$$\eta_{EL} = \eta_R \eta_E \eta_O \qquad (3.3)$$

where η_O is the optical extraction efficiency for light from the LED chip. In an indirect gap material η_O may be $\gtrsim 25\%$; it is ~25% for yellow-green near-gap light in GaP:N and ~60% for the much more weakly absorbed red light in GaP:Zn,O. Thus, η_R is ~30% for the best GaP:Zn,O devices quoted in Table 3.1, but is only ~3% for the best GaP:N. The Zn,O activator in p-type GaP is a very exceptional case. Effectively, luminous equivalence of the emitted light has been traded for large trap depth of the minority electron, here ~0.3 eV. This large depth E_e ensures that captured minority electrons have a good chance of hole capture and radiative recombination before thermal re-emission can occur. The probability of the latter process contains the usual factor $\exp(-E_e/kT)$. It is therefore very large for more typical trap depths of a few tens of meV found in GaP:N, $GaAs_{1-x}P_x$:N and the direct gap LEDs. Relatively small concentrations of the deep centers usually present inadvertently can compete very effectively with the shallow activators such as N because of the much greater thermal re-emission rates of the activators. The recombinations at these deep states need not be radiative. In fact we shall see in Section 3.6 that they are frequently totally non-radiative. In either event, they provide an unwanted competitive decay channel for the minority carriers injected into an LED and give rise to an aptly named shunt path in the recombination kinetics. Unfortunately, the competitive effect of deep centers with but modest minority carrier cross sections is augmented by the thermal emission effect just mentioned, to the point where most of the minority carrier recombinations occur through the shunt path in practical material. A further unfortunate circumstance is the fact that the rather small cross sections typical of very deep centers can be greatly augmented by the effects of large lattice deformation which accompanies efficient energy loss to large numbers of lattice phonons, as discussed in Section 3.7.

It is also true that non-radiative processes may be significant at the luminescence activators themselves, particularly if the thermal emission rates are

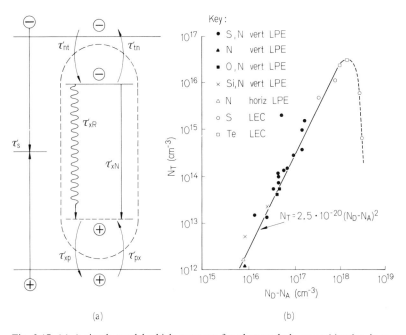

(a)   (b)

Fig. 3.17. (a) A simple model which accounts for electron-hole recombination in *n*- and *p*-type GaP: N. The transitions through the N isoelectronic trap involve the indicated electron and hole capture and thermal emission times, the radiative lifetime τ_{xR} and the non-radiative Auger lifetime τ_{xN}. It is necessary to invoke parallel competing recombinations through at least one unidentified center, which is hypothesized to involve an energy level far from either band edge. The lifetime τ_s for these "shunt path" recombinations dominates both minority carrier lifetimes in GaP [3.49].

(b) The increase in concentration of an unidentified trap 0.42 eV below the conduction band with electron concentration in *n*-type GaP. This trap is observed in thermally stimulated current spectra for a wide variety of donors in both LEC and LPE GaP and may be a dominant minority carrier recombination center in *p*-type material [3.58]

not too great. As an example, about 50% of the recombinations of electrons bound to Zn,O in optimally doped red GaP LEDs are of the non-radiative Auger type [3.48], involving at least one free hole. Free carrier Auger processes also reduce η_{EL}, the minority carrier lifetime τ_{mc} and diffusion length L at sufficiently high majority carrier concentrations in the near-gap emitters, for example apparently above $\sim 10^{18}$ cm^{-3} in both p- and n-type GaP [3.49, 50].

However, by far the most severe limitation on the performance of all the near-gap emitters at optimum doping levels is provided by the shunt path recombinations through centers which remain unidentified to this day. The optical studies described in Section 3.2 have been sufficiently thorough for the mainstream LED hosts shown in Table 3.1 that we are now unlikely to discover better luminescence activators. We can only hope for further improvements in performance through a reduction in the concentration of the principal shunt path centers. This has proved a difficult job in the absence of more evidence on their nature than we currently possess. For this reason, there is now much interest among the principal LED manufacturers in the exploitation of the techniques described in the Section 3.6 for the characterization of deep traps, painstaking as they are compared with the study of the luminescence activators by optical spectroscopy.

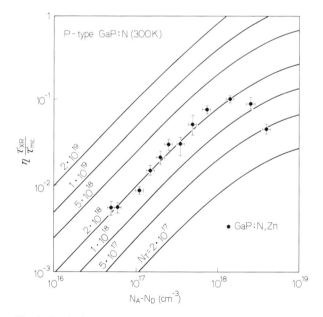

Fig. 3.18. The increase in the product $\eta\tau_{xR}/\tau_{mc}$ with hole concentration in p-type LPE GaP: N. The linear increase follows from the kinetic model of Fig. 3.17a up to the point where the Auger recombination rate τ_{xN} becomes significant. The lines are theoretical and fit the experimental data for a nitrogen concentration N_T which agrees with independent estimates from optical absorption [3.49]

The simplified kinetic model for GaP:N shown in Fig. 3.17a contains the essential elements for the analysis of a typical near-gap emitter. Effectively, we can write

$$\eta_{\text{EL}} \propto \eta_{\text{R}} \propto \tau_{\text{mc}}^{\text{e}}(N)p \tag{3.4}$$

where we consider only p-type material for illustration, minority carrier lifetime $\tau_{\text{mc}}^{\text{e}}$ and bulk radiative efficiency η_{R} and (N) is the activator concentration. The photoluminescence lifetime $\tau_{\text{s}}^{\text{PL}}$, often a convenient diagnostic, is essentially equal to $\tau_{\text{mc}}^{\text{e}}$ for GaP:N. Thus $\tau_{\text{mc}}^{\text{e}}$, which defines η_{EL}, is set by shunt path recombinations. This is true for near gap light with or without N doping in both GaP and indirect $GaAs_{1-x}P_x$, although the *internal* efficiency may be much larger in direct gap $GaAs_{1-x}P_x$, rendering (3.4) inappropriate. However, this relationship has been checked in green GaP up to the limiting p for the onset of Auger processes (Fig. 3.18). The dependence of η_{EL} on τ_{mc} within controlled batches of material has been verified by several groups [3.52]. Measurements of τ^{PL} are subject to a number of serious pitfalls if such bulk material relationships are sought and precautions must be taken to avoid them [3.52].

3.6 Techniques for the Detection and Characterization of Deep Levels

We are interested in the labeling and, if possible, the identification of the major centers responsible for the shunt path recombinations, at least in the most important LED materials listed in Table 3.1. We realize that the shunt path recombinations may not be totally non-radiative. In favorable circumstances, recombinations at even predominantly non-radiative centers can still be analyzed most conveniently through optical spectra, as was possible for Auger-dominated bound exciton recombinations involving neutral donors and acceptors in GaP and Si [3.53, 54]. However, they are likely to involve very deep traps for reasons discussed in Section 3.5. We shall see that electron-hole recombinations at deep traps are likely to show strong lattice coupling. This means that the limited proportion of recombinations which are radiative is likely to produce photons with a broad distribution in energies, making identification and even detection still more difficult. Most of the diagnostic work has been performed on GaP to date. Many weak, broad luminescence spectra are known. The problem is to find which if any are related to principal shunt paths. In the most relevant recent work, *Bachrach* et al. [3.55] made a study of GaP:Te,N and GaP:S,N, selected on the grounds that they find like others low energy broad luminescence bands specific to these dopants. There has been a particularly large amount of speculation about deep states involving Te complexes [3.56, 57]. The inset in Fig. 3.19 shows some kinetic aspects of broad band luminescence in GaP:Te,N. The radiative efficiency of

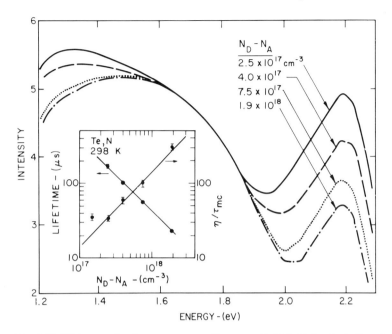

Fig. 3.19. Wide range luminescence spectra for *n*-type GaP:Te,N for the indicated electron concentrations. The intensity is recorded on a logarithmic scale. The narrow band near 2.2 eV is dominated by N-induced exciton recombinations. There are two poorly resolved broad bands at low energies. The higher energy band peaking near 1.55 eV has lifetime and normalized efficiency which vary with electron concentration as shown in the inset. It may involve the 0.42 eV electron trap discussed in Fig. 3.17 (b) [3.55]

the center increases linearly with $N_D - N_A$. The centre is also enhanced by the chemical conditions used for N doping, although it does not involve N directly. The high energy tail of the broad luminescence band in Fig. 3.19 suggests a no-phonon transition near ~1.85 eV, which could correspond to a free to bound recombination at an electron trap of depth ~0.4 eV below E_c. This could be the center whose concentration varies as $(N_D - N_A)^2$ reported in thermally stimulated current studies of *n*-type LPE GaP [3.58] (Fig. 3.17b). However, the luminescence studies show that the recombination rate through this center is low. It is easily saturated at normal LED drive levels, though at low drive levels it accounts for ~40% of the total electron-hole recombinations only ~3% of which are radiative [3.55].

We must therefore conclude that at relevant LED drive levels τ_{mc} in GaP is dominated by recombinations which provide insignificant luminescence, at least down to ~1.2 eV. We therefore require diagnostic techniques that show up traps by non-radiative means. The photocapacitance technique is eminently suitable. The deep traps are detected by the change in the capacitance of a depletion layer when the electronic occupancy of traps within the layer is changed through optical excitation. The depletion layer may be contained in

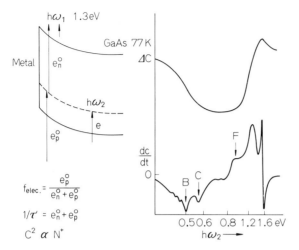

Fig. 3.20. To the left is a schematic view of a Schottky barrier in an *n*-type semiconductor with a single deep trap, shown dashed. The fractional occupancy of electrons in this trap f_{elec} and the characteristic response time of this population to changes in the intensity of the bias light of energy $\hbar\omega_1$ just below E_g are defined by the optical transition rates e_p^o and e_n^o as shown. Changes in f_{elec} can be detected by changes in the junction capacitance C. The spectra of VPE GaAs on the right show the benefit in recording dc/dt (effectively $dc/d\hbar\omega_2$) as a function of $\hbar\omega_2$ rather than just ΔC in the typical situation in an undoped semiconductor where several deep traps appear together. The polarity of the structure in this double source differential photocapacitance (DSDP) spectrum distinguishes e_p from the complementary transition (CT) e_n and is negative for e_p. Peak *F* has been attributed to Cr_{Ga} [3.66]

a Schottky barrier (Fig. 3.20) or a *pn* junction. Profile information may be obtained using changes in depletion layer width through appropriate control of the externally applied bias voltage. A very convenient feature is the direct measure of trap concentration provided from the changes in junction capacitance. A variety of capacitance techniques was first applied to the study of the traditional semiconductors Ge and, particularly Si [3.59]. The first thorough application of photocapacitance to wider gap semiconductors was a study of the deep O_P donor in GaP [3.60]. This study confirmed the binding energy for a single electron obtained from earlier optical spectra [3.61] and provided reliable measurements of concentrations and optical cross sections for the first time. A surprise, confirmed in later work [3.62] although alternative ideas exist [3.63], was the abnormal behavior for binding a second electron. It is suggested that a very large degree of lattice relaxation causes the sum of the complimentary transitions, CT = valence→level + level→conduction band, to exceed E_g by a large margin. This gives the second electron greater binding energy than the first! Measurement of the change in the trapped charge population during pulse bias of variable time duration provides a measure of both majority and minority carrier capture cross sections [3.64], two most important parameters. The bias conditions are adjusted so that these processes are always measured in neutral material beyond the depletion layer, the relevant situation. This is an important point, since carrier capture and emission rates are generally greatly perturbed by typical electric fields well within a depletion layer [3.65]. Effects due to the diffusion of the minority carrier population beyond the "edge" of the depletion layer must also be allowed for [3.65]. Knowledge of these capture cross sections, the recombination lifetime and the concentration for a given center allows a secure judgment to be made on its contribution to recombination, non-radiative or not. In this way it has been found that the isolated O_p donor is not a dominant recombination center, even in p-type GaP where its concentration is greatest [3.48]. In fact, there is more chance of an O-dominated shunt path in n-type GaP, since the cross section of the second electron state is greatly enhanced by the strong lattice relaxation it exhibits [3.62]. That is, the 300 K properties of this system are much nearer to the high temperature limit described in Section 3.7. However, careful growth techniques which depress [O] far below even the low value typical for n-type LPE GaP show no effect on the minority carrier lifetime and do not correlate with the 0.75 eV trap described in Section 3.8 [3.89].

The photocapacitance experiments of *Kukimoto* et al. [3.60] are tedious and not well suited to the initial survey of levels present in large batches of material or to the study of several levels present together in a single sample. Two methods have been developed to deal with this frequently occurring situation. We will first describe the recently developed double source differential photocapacitance (DSDP) method [3.66]. The basic trouble with conventional single source photocapacitance is that the capacitance response time to changes in the necessarily weak probe light from a scanning monochromator are typically uncomfortably long. Reliable data can only be collected over time scales long compared with these times. The establishment of a single spectrum then becomes an inconveniently lengthy task, with consequent problems of detection sensitivity and instrument instability. A solution to this problem is given in Fig. 3.20a. The bias light not only defines the electron occupation probability in each deep level at some steady value, conveniently neither 0 nor 1 in general, but also sets to an acceptably short value the relaxation time against further small changes imposed by the relatively weak probe light. The system is then in optical equilibrium, as can be checked by reversal of the scan direction of the probe light energy. It is only convenient to cover the energy range between ~ 0.3 eV from either band edge by this technique. Apart from increasing difficulties with the optical source and atmospheric absorptions, traps of shallower depth begin to have inconveniently short carrier release rates, even at very low temperatures, because of the influence of the junction electric field [3.65]. Application of this technique to GaAs has revealed the structure shown in Fig. 3.20b in typical undoped VPE material in a scan time of a few minutes. Since several traps with close-spaced levels frequently appear together, they are best revealed by differentiation of the photocapacitance signal. Initial results of this study have shown that, as might have been expected, the Lucovsky model [3.67] is not well suited to the description of the energy dependence of the absorption cross section of very deep traps in semiconductors whose band structures are poorly approximated by spherical functions over the relevant energy ranges [3.68]. The CT energy sums obtained from peaks in the DSDP spectra are significantly larger than the direct bandgaps of materials like GaAs with low density of states conduction band minima, for reasons other than lattice relaxation [3.62]. The principal limitations of the DSDP method are that it will not be sensitive for centers with abnormally low optical cross sections and that it gives no direct information on capture cross sections. However, it is a very useful survey technique which ideally should be complemented by the deep level transient spectroscopy (DLTS) method [3.69] described next. The DSDP method is particularly suited to the study of the very deep traps found in the wider bandgap semiconductors, some of which cannot be reproducibly cycled to the high temperatures then required by the DLTS method. Apparently, some centers are best detected by DSDP, for example Cr in GaAs, and some by DLTS, for example the level often attributed to O in GaAs as well as the important 0.75 eV hole trap in GaP described in Section 3.8. The DLTS method does have the drawback of only detecting majority carrier traps using Schottky barriers, like the thermally stimulated current technique used for Fig. 3.17b. However, it may be possible to observe

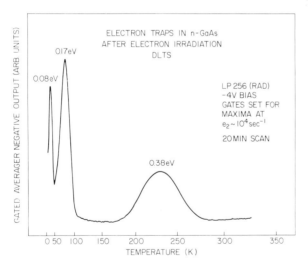

Fig. 3.21. An example of deep level transient spectroscopy (DLTS) (Fig. 3.22) in *n*-type LPE GaAs showing three majority carrier traps in the upper half of the bandgap produced by electron irradiation. In contrast to the DSDP technique (Fig. 3.20), the DLTS method detects minority carrier traps only if a p^+n junction is prepared and the short DC pulse which sets the trap occupancy for subsequent thermal emission under reverse bias takes the junction sufficiently far into forward bias to saturate the minority carrier traps. The spectrum shown here demonstrates the ability of the DLTS method to detect large thermal emission rates which can reveal the presence of shallow traps well overlapping the energy range of major application of the photoluminescence technique (Fig. 3.2) [3.69]

the same center in material of opposite conductivity type, perhaps most easily achieved in a diffused *pn* junction structure. If an asymmetric *pn* junction is prepared to ensure the study of traps in a defined type of material, minority carrier traps may be filled by saturation injection under appropriate forward bias in the DLTS method [3.69].

The DLTS method makes use of thermal rather than optical release of charge from traps in a depletion layer [3.69]. It is also well suited to material survey, particularly where several different traps appear. The novel feature of the DLTS technique is the use of a short DC bias pulse to fill the traps, with fast detection of the subsequent decay of non-equilibrium excess charge after removal of the reduced reverse bias or forward bias. The tuned rate window is capable of operating at rapid release rates (Fig. 3.21). It is then possible to detect traps as shallow as 0.1 eV or even less, thus covering the energy range within the bandgap shown in Fig. 3.2. Trap detection occurs when the thermal release rate for a given trap becomes equal to the preset rate window as the device temperature is slowly increased (Fig. 3.22). The trap depth is determined from a measurement of changes in the temperatures of peak capacitance response induced by changes in the rate window settings. Capture cross sections are measured in the neutral material during the DC forward bias pulse, as described earlier. Variants of the DLTS method are possible, in which the apparatus and the extraction of data may be both simpler and more efficient. However, *Lang*

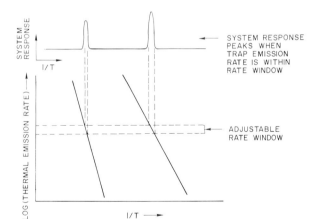

BASIC IDEA OF DLTS METHOD

Fig. 3.22. Demonstrates the basic principle of the DLTS technique used to obtain the spectrum in Fig. 3.21. The lower part contains typical activation energy plots for the electron populations in two traps of significantly different depth. The upper part illustrates schematically the response of an electronic detection system tuned to a particular electron emission rate determined from the junction capacitance. Traps of difference depth produce the set emission rate at clearly distinguishable temperatures [3.69]

and co-workers have determined the properties of many deep traps in GaAs and GaP using a double boxcar to define the rate window [3.62].

3.7 Carrier Capture by Multiple Phonon Emission and Recombination-Induced Defect Motion

Strong support for these two important processes has emerged from the early applications of the DLTS method to the study of deep traps. Measurement of the capture cross sections of these traps in GaAs and GaP using a double pulse mode of operation [3.69] has shown a strong variation with temperature for most of them. Many of the capture cross sections increase rapidly with temperature. Levels with the smallest cross sections at low to moderate temperature show the greatest rates of increase at higher temperature (Fig. 3.23). The general consequence is that the electron and hole capture cross sections for many different traps extrapolate to a common narrow range between 10^{-15} and 10^{-14} cm^{+2} in the high temperature limit, whether or not they possess a large temperature dependence at lower temperatures. Thus

$$\sigma_T = \sigma_\infty \exp\left(-E_\infty/kT\right) \tag{3.5}$$

where E_∞ lies in the range 0 to ~ 0.56 eV, that is it may be significant compared

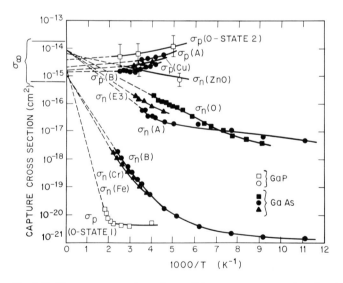

Fig. 3.23. The temperature dependence of the capture cross sections for electrons, subscript n and holes, subscript p for the indicated traps in GaAs and GaP. The common high temperature limit is denoted by σ_∞. The low temperature plateau is determined by the detailed form of the local lattice readjustment consequent upon carrier capture for each particular center [3.62]

with the level depth, and is approximately equal to E_B in Fig. 3.24. Henry and Lang note that such strong temperature dependence is a hallmark of energy loss to multiphonon emission (MPE) during carrier capture. In fact MPE can readily account for the very large capture cross sections possessed by just those deep states which may predominate in the carrier capture and recombination processes and therefore control the shunt path of Fig. 3.17a (Sec. 3.5).

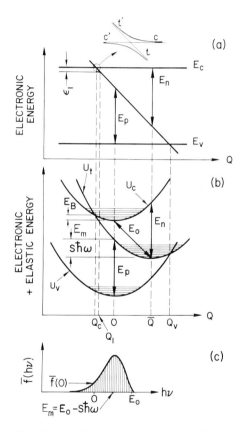

Fig. 3.24a–c. The variation of the electronic energy (a) and the electronic plus elastic energy (b) of a localized center with strong coupling between the electronic energy and the lattice, parameterized in terms of a single local lattice coordinate Q from a zero set when the trap is empty. There is an anticrossing between the conduction band E_c and the trap energy at Q_c, as shown at the top. The adiabatic approximation in which the lattice frequency is no longer negligible compared with the electronic frequency breaks down at Q_1 near Q_c and at a corresponding position near the valence band anticrossing at Q_v. The radiative capture line shape is shown in (c) with a spectral peak E_m at $S\hbar\omega$ below the no-phonon threshold E_0 and a value $\bar{f}(0)$ at $h\nu = 0$. For linear electron-phonon coupling, the peak in the photoexcitation spectrum is at $E_n = E_0 + S\hbar\omega$, where $S\hbar\omega$ is the increase in electric binding energy on lattice relaxation [3.62]

The plateau values of the cross section shown in Fig. 3.23 are given by

$$\sigma = \frac{C}{N\langle v\rangle}\, \exp\,(-\gamma E_m/\hbar\omega) \qquad (3.6)$$

where C may be $\sim 10^9\ \mathrm{s}^{-1}$, N is the number of traps and $\langle v\rangle$ an average carrier velocity before non-radiative capture. The factor γ can be evaluated exactly if the dominant lattice frequency w is independent of the electronic state [3.70]. Then,

$$\gamma = \ln\,[E_0/(E_0 - E_m)] - E_m/E_0 \qquad (3.7)$$

where E_0 and E_m are shown on the configurational coordinate diagram in Fig. 3.24. This factor is probably ~ 1 for centers with moderately strong coupling in semiconductors, but is ~ 2–3 for very weakly coupled rare earth ions. Since σ in (3.6) decreases exponentially with the number n of phonons to be emitted, $n = E_m/\hbar\omega$, it is easy to see why the plateau values in Fig. 3.23 should depend so sensitively on the detailed nature of the center. It also follows that capture must become radiative for sufficiently large n. The limiting value is $n \sim 5$ for rare earths, where the radiative probabilities and phonon coupling are both relatively small [3.71]. As yet, there has been no correspondingly systematic study for deep centers in semiconductors. However, it is known that electron capture at the deep donor O_P, which exhibits weak phonon coupling in its optical spectra, is predominantly radiative at $n \sim 15$ [3.61]. As temperature is increased, vibrational excitations in the initial electronic state introduce a temperature-dependent term in the capture cross section. This can be derived analytically for the weak interaction limit and is represented by the additional multiplicative factor $[1/1 - \exp\,(-\hbar\omega/kT)]^n$. The activated form of (3.5) is reproduced only at high temperatures such that $kT > E_\infty/\gamma n$. The important point here is that only in the high temperature limit can σ be sufficiently large to dominate carrier capture and recombination for the concentrations of deep centers which occur inadvertently in the best device grade semiconductor material. Thus, the centers which dominate the shunt path will be those which are in or close to this limit at normal device operating temperatures, just above 300 K. We see in Fig. 3.23 that electron capture at O_P is well below this limit, the principal reason why this center does not dominate the shunt path despite the relatively high concentration of O_P in p-type GaP:Zn,O [3.48]. Thus, centers with relatively low E_∞ and therefore large lattice relaxation between initial and final electronic states (Fig. 3.24) must predominate unless those with smaller σ are favored by being present in much larger concentration.

The other plausible method for the non-radiative dissipation of the large energies released on carrier capture and recombination in a wide gap semiconductor is the Auger effect, known to be important for near gap bound excitons in certain cases [3.53, 54]. However, *Henry* and *Lang* note that the free carrier Auger cross sections are rather small at relevant carrier concentrations, for example $\sim 10^{-18}$ cm^{+2} for processes involving free holes and the Zn-0 bound exciton in GaP for 10^{18} cm^{-3} holes at 300 K [3.72]. It is true that the Auger rate increases dramatically with binding energy [3.54]. However, in a careful study of a persistant deep level in undoped LPE GaAs, *Henry* and *Lang* failed to find the linear increase of capture cross section with equilibrium carrier concentration expected for capture dominated by the Auger process. They conclude that the Auger process is not responsible for the thermal activation of the capture cross section observed for this trap. Of course, other traps may exhibit this effect, perhaps with greater probability for intermediate-depth traps with relatively modest electron-phonon coupling, especially in indirect gap semiconductors where the Auger process is more significant compared with radiative recombination [3.73]. However, there is no present evidence for the dominance of this process for any *deep center* recombinations, which are also much more likely to exhibit the large local lattice relaxation on carrier capture necessary for the MPE process. An alternative, widely discussed energy loss mechanism presupposes a series of real electronic states through which capture can occur with sequential energy liberation of a small number of phonons, perhaps just one at each step [3.3]. This cascade process is responsible for the large *decrease* in σ with increasing temperature for capture through shallow donor excited states observed at very low temperatures [3.74]. The problem for very deep centers is that such ladders of intermediate levels are neither expected nor found. Neutral centers usually do not have excited states, while the excited (Coulomb) states of charged centers all lie within one or two optical phonon energies of the band edges. There is a possibility that one or more relatively deep excited electronic states could occur for a center which undergoes strong asymmetric distortion upon capture of a carrier, the case of the static Jahn-Teller effect [3.74a]. However, no well-authenticated example of such a non-radiative center is yet identified. In any case, this is just a special example of a center with strong phonon coupling, where MPE may be strong even in the absence of additional electronic states. The consequence of this is that capture into the ground state of a deep charged center which does not show strong lattice coupling and therefore the MPE, like the first electron bound at O$_p$ in GaP, can only be *radiative* at low doping levels.

Inter-center Auger (or cross-relaxation) processes may take over at appropriate joint impurity densities [3.61].

The relation between emission rate, trap depth and capture cross section given from detailed balance is

$$e = \sigma \langle v \rangle N_c \exp(-\Delta E/kT)/g \qquad (3.8)$$

where N_c is the density of band states and g is the level degeneracy. Eqs. (3.5) and (3.8) show that the true activation energy ΔE is the difference between the values measured in carrier emission and carrier capture. If $\sigma \langle v \rangle$ is a strong exponential function of temperature as for σ_n(Cr) in Fig. 3.23, this temperature dependence produces a large correction to the directly measured activation energy. *Henry* and *Lang* have developed a simple theory for MPE in which vibrations of a single lattice coordinate modulate the potential well depth for binding at a deep center [3.62] (Fig. 3.24). The MPE process is strong if there is a large electron-phonon coupling, marked by a large change in equilibrium lattice coordinate and energy of the electronic state during capture of an electronic particle. This situation is qualitatively very familiar in strongly ionic lattices, for example the well known *F* center in the alkali halides [3.75]. The energy change is dissipated by the emission of the appropriate number of phonons into the lattice at large. Immediately after recombination, the energy resides in lattice vibrations of local character. A high proportion of the vibrational kinetic energy of these modes involves modulation of the local lattice coordinate, as shown in Fig. 3.24.

In favorable cases the large adiabatic energy release to the lattice consequent from MPE can induce a permanent change in the atomic arrangement near the defect, resulting in the appearance of a recombination-enhanced annealing effect. This is the second of the important results which have recently emerged from the systematic study of deep levels by the DLTS technique. *Lang* and *Kimerling* [3.76] have studied the effect in greatest detail for a 0.31 eV electron trap observed in electron-irradiated GaAs (Fig. 3.25). The thermal quenching rate for this center is 1.4 eV. This is reduced to only ~ 0.34 eV if thermal annealing is carried out under forward bias currents of ~ 10 A cm^{-2}. We note in Fig. 3.25b that the energy released during electron-hole recombination at this center, ~ 1.1 eV, is essentially identical to the difference in thermal activation energies $E_\mathrm{T} - E_\mathrm{CE}$. This strongly suggests that the recombination energy contributes to the atomic rearrangement involved in the annealing of this defect, passing through an intermediate stage of violent local lattice vibration. This is a particular example of the Gold-Weisberg "phonon-kick" model [3.77] of defect formation, which has been the subject of much speculation. Recent circumstantial evidence includes the observation of a rapid increase of degradation rate with the energy gap within a series of semiconductor alloys [3.78]. However, the data shown in Fig. 3.25 probably provide the most definitive evidence for this mechanism available to date.

The theory of capture cross sections of *Henry* and *Lang* [3.62] represents a development from previous work in which the large enhancement related to the breakdown of the adiabatic approximation is discussed in a way which emphasizes physical insight. In the adiabatic approximation, the electronic transitions are regarded as occurring in a frozen lattice, indepedent of the value of the instantaneous lattice coordinate Q in Fig. 3.24. The approximation breaks down at Q_1 where the free carrier and trap electronic states come together and the transition energy is no longer large compared with the lattice vibrational energy. It is just this extreme range of Q over which the lattice spends a relatively large amount of time and the free and bound carrier states are mixed by the electron-lattice coupling, measured from the change in the potential well depth of the center for a given lattice displacement. The free to

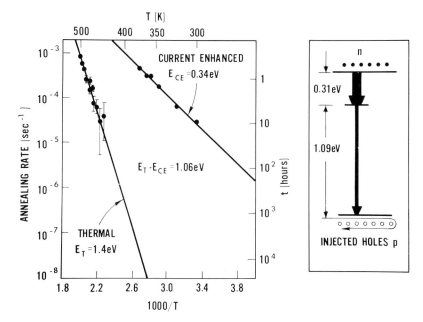

Fig. 3.25. To the left are the annealing rates observed in the concentration of a trap 0.31 eV below the conduction band in LPE GaAs damaged by electron irradiation, measured without and with a forward bias current in the P^+n diode used in this DLTS study. The current-enhanced annealing rate was measured under saturation conditions at 2.5 A cm^{-2}. The local energy available on re-combination of injected minority carriers shown to the right is approximately identical to the reduction in the thermal activation energy for annealing this defect obtained from the difference in the Arrhenius plots on the left [3.76]

bound state transition occurs in this range of Q. The energy E_∞ in (3.5) is essentially the energy difference in the free particle energy contour on the configuration coordinate diagram between the equilibrium lattice coordinate and the critical value for crossover of the bound and free states. *Struck* and *Fonger* [3.79] have recently pointed out that in certain circumstances interstate non-radiative transitions may be significant over a wide range of lattice co-ordinates below the crossover. This can cause the capture cross section to be rather poorly represented by a single exponential form. Henry and Lang show that $\sigma = Af(0)$ where $f(hv)$ is the normalized line shape for radiative capture [3.62]. The radiative rate is zero at $hv = 0$ because of the familiar $(hv)^3$ factor it contains. The coefficient A is insensitive to the detailed properties of the host and impurity, while $f(0)$ is determined by the relaxation properties of a particular center in a particular host. The value of σ_∞ for a neutral center is estimated at $\sim 1 \times 10^{-15}$ cm^{+2}, with a 10–20-fold enhancement factor for a Coulomb attractive center. Little Coulomb enhancement is expected for a very deep donor or acceptor. These values of σ_∞ are consistent with the observed limiting range in Fig. 3.23. This agreement gives strong support to the conclusion that capture by MPE is a process of rather wide generality at deep states, at least for the important III-V semiconductors GaAs and GaP.

3.8 Deep Levels and the Shunt Path in GaP

Although there has been much work on the role of macroscopic defects in the degradation of GaAs LEDs and injection lasers, described briefly in Section 3.11, most of the work on dominant microscopic luminescence killer centers has been carried out in GaP [3.80]. A considerable number of energy levels have been located by the TSC technique. A few involve known impurities or defects which can be avoided, for example those at 0.39 eV and 0.55 eV above the valence band associated with Cu [3.81]. Of the levels determined by this method, only the unidentified electron trap 0.42 eV below the conduction band (Fig. 3.17b) has a sufficiently persistent density and large capture cross section, $\sim 7 \times 10^{-15}$ cm^2, to make it a potentially dominant minority carrier recombination center, here for electrons in p-type GaP. However, the most important center detected to date has a level 0.75 eV ebove E_V. *Hamilton* et al. [3.82] observed the level by filling it with majority carriers in n-type GaP on forward bias of a Schottky barrier. Minority carriers generated by optical interband excitation can then be captured under reverse bias and subsequently released by thermal ionization. The trap depth is determined from the thermal release rate, while the minority carrier cross section $\sim 2 \times 10^{-14}$ cm^{+2} is determined from the capture rate under optical excitation. The optical cross sections of this center are negligible, so it cannot be detected in photocapacitance. The large minority carrier cross section and significant concentration $\sim 10^{12} - 10^{14}$ cm^{-3} for the 0.75 eV trap suggest that it should be a very important recombination center. This prediction has now been verified in several sets of n-type LPE samples, where τ_{mc}^p has been shown to correlate inversely with [N$_{0.75}$]. Particularly important is a recent suggestion that the correlation may extend to VPE GaP, which normally exhibits about an order lower τ_{mc}^p compared with LPE. The form of the I, V characteristics of LPE GaP LEDs [3.47] shows that the dominant recombination in the depletion layer involves a trap ~ 1 eV above the valence band, in reasonable agreement with these capacitance data. Thus far, there has been no determination of the dominant shunt path center in p-type GaP. It might be the 0.42 eV trap of Fig. 3.17b or even the 0.75 eV trap again, since the electron capture cross section of the latter has not been measured.

Little is known of the nature of the 0.75 eV center. Appeal to theory for guidance on the type of centers which may possess these properties is not very fruitful at present. There are only some rather general ideas that centers containing vacancies may possess the large lattice relaxation between different charge states necessary for efficient energy loss by MPE (Sec. 3.7) because of the local softening of the lattice they produce. The simple idea that the large capture cross section and non-radiative properties of this 0.75 eV trap in GaP are due to MPE, with the center near the high temperature limit of Fig. 3.23, seems possible only if E_∞ is very small for this center and the electron-phonon interaction is very nonlinear, since σ_h is relatively temperature insensitive between 77 K and 300 K [3.82a]. *Watkins* [3.83] has noted that cation vacancies frequently exhibit large static Jahn-Teller distortions in Group IV, II-VI and I-VII semiconductors and predicts similar behavior for III-V semiconductors. The critical factor may be the presence of unfilled *p*-orbitals on the atoms surrounding a cation vacancy, not present for anion vacancies. Many complex centers containing vacancies have been assigned to broad recombination bands, particularly in GaAs. However, the assignments are based upon circumstantial evidence of a kind which has been found very misleading in some cases [3.3, 84]. It may well be that these vacancy-containing centers form an important general class of luminescence killer centers. The most useful clue to the nature of the 0.75 eV trap is the correlation with dislocation density shown in Fig. 3.26.

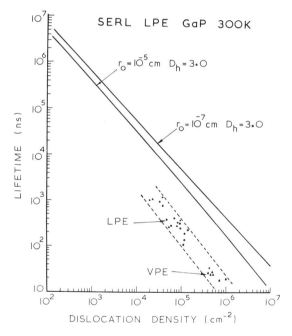

Fig. 3.26. The points show the variation of minority carrier lifetime observed in LPE and VPE GaP with dislocation density measured at the same positions of the epitaxial layers as the lifetime. The solid lines show the limits for recombination through the dislocation cores based upon a two-dimensional diffusion model with the indicated diffusivities and dislocation core radii within which the non-radiative minority carrier recombination rate is assumed to be infinite. The lifetimes also correlate with the density of a trap 0.75 eV above the valence band measured between a set of carefully selected LPE and VPE layers with specially uniform properties. More recent data for iso-thermally grown LPE GaP fall approximately midway between the upper dashed line and the lower solid lines and are less scattered [3.85]

There is some dispute about the values of τ_{mc}^p for injection levels comparable with commercial LEDs [3.28]. *Blenkinsop* et al. [3.85] showed that the measured lifetimes in their LPE and VPE GaP fall below values calculated for minority carrier recombination through the dislocation cores, by a factor somewhat greater than 10, assuming the reasonable values of core radii and diffusivity in Fig. 3.26. They believe that the shunt path is dominantly due to recombinations occurring between the dislocations, involving the unknown center with 0.75 eV energy level which does not appear in transmission or scanning electron microscope TEM and SEM surveys (next section). This view is consistent with the general belief that dislocation recombination reduces the overall efficiency of good quality LPE GaP by only about 10% [3.86]. Quite high dislocation densities may be necessary before the average separation becomes comparable to the minority carrier diffusion length when dislocations dominate the shunt path, for example $[N_d] \gtrsim 5 \times 10^6$ cm^{-2} in GaAs at 300 K [3.87] but only $\gtrsim 10^5$ cm^{-2} in high quality LPE GaP [3.88]. Using a one-dimensional model for dislocation recombination [3.88], *Kuijpers* et al. [3.28] estimated that τ_{mc}^p *is* limited by this process for a variety of LPE GaP and VPE GaP and GaAs$_{1-x}$P$_x$ with $x \gtrsim 0.8$, provided that τ_{mc}^p is measured at sufficiently high excitation densities. These workers found a much greater variation of τ_{mc}^p with injection level for VPE than *Young* and *Wight* [3.52]. These conflicting opinions result partly from an underestimate of τ_{mc}^p at a given $[N_d]$ from the one-dimensional model and partly from a uniform 2–3 times larger experimental τ_{mc}^p reported by *Kuijpers* et al. [3.28]. The latter discrepancy is removed if more recent data obtained from a set of isothermally grown LPE GaP are considered [3.89]. The present concordant view is that τ_{mc}^p can be limited by dislocation core recombinations in state of the art GaP, LPE and perhaps VPE. However, further reduction, usually correlated with the 0.75 eV microscopic defect, is a common property of average quality material [3.89].

3.9 Dislocations and Luminescence Topography

As in GaAs and other semiconductors, it is easy to demonstrate the efficient luminescence killer action of the dislocation cores in GaP through SEM topographs (Fig. 3.27a). The dislocations may be revealed most easily by etching. A given etch may not reveal all dislocations, so there are frequently more dark spots than dislocations. However, electrolytic etch techniques may be used to prepare foils of bulk semiconductor sufficiently thin for study in the high voltage TEM. Then, dislocations of different type can be revealed through the contrast imposed by their local strain fields in different geometries of electron diffraction. The data in Fig. 3.27 are part of a correlation in which each of 17 dark spots in a given region of a piece of LPE GaP was related to a dislocation which was then fully characterized [3.90]. Dislocations of all types were found, edge, screw and intermediate. All gave dark spots of similar contrast in the SEM topograph (Fig. 3.27a). The recent observations of *Heinke*

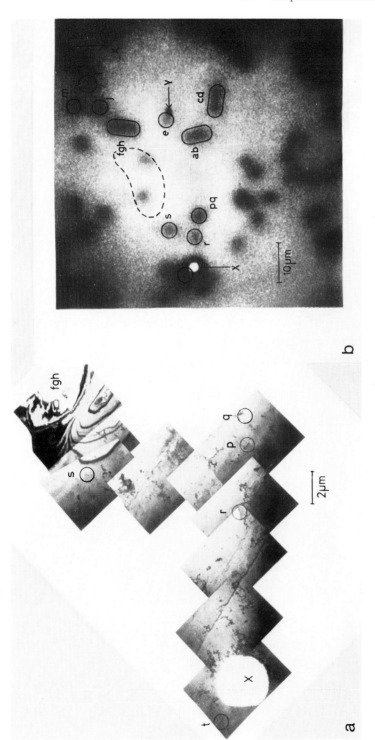

Fig. 3.27. (a) Shows a montage of transmission electron microscope topographs from a region of the LPE GaP crystal used for the cathodoluminescence topograph in (b). Dislocations are revealed in (a) through the effect of their local strain field on electron diffraction and in (b) through the local excess non-radiative recombinations they induce. Every linear dislocation line image in (a), labeled $p \rightarrow t$ is matched by a dark patch in the corresponding portion of (b). The white spot at X is a hole in the epitaxial layer, while region fgh is a larger hole produced in the foil during the thinning necessary for transmission electron topography [3.89, 90]

and *Queisser* [3.183] in GaAs are of interest here. They found that dislocations introduced by controlled plastic deformation at 580 °C have much stronger non-radiative recombination effects than those introduced during crystal growth. Much of this excess activity can be annealed away above 600 °C. Despite evidence from the simultaneous enhancement of broad-band luminescence attributed to vacancy-impurity complexes, the causes of these annealing effects are a subject of speculation at present. Much more work will be necessary before these matters are clarified.

It must be recognized here that there are likely to be several significant shunt path centers, just as there are several efficient luminescence activators even in a single host lattice. The 0.75 eV center seems to have wide significance in n-type GaP, as do the dislocation cores [3.91]. The dislocation density in epitaxial GaP is usually controlled by that in the substrate. This is a particular problem in VPE GaP, since the dislocation density in most available LEC substrate material is relatively high and there is no analogue of the interrupted growth technique used in LPE GaP to reduce the density of propagating dislocations [3.92]. The dislocation density can be raised well above that of the substrate by injudicious growth procedures, for example if the concentration of the N activator in VPE GaP is increased too far above the solubility limit for equilibrium growth. The lattice mismatch can be relieved by jogging interfacial dislocations up to $\sim 5 \times 10^{18}$ cm^{-3} [N] [3.93]. A very large increase of inclined dislocations sets in for [N] $\gtrsim 10^{19}$ cm^{-3}, accompanied by a marked decline in overall luminescence performance.

At present, there is no evidence for specific energy levels attributable to non-radiative recombination through dislocation cores. The question of their possible existence is an interesting one. *Benoit à la Guillaume* [3.94] gave some early evidence for specific broad luminescence bands in Ge, attributed to recombinations through dislocations including some produced by controlled plastic deformation. It has not been proved that this luminescence is spatially located at the dislocation cores. We have seen that the 0.75 eV shunt path trap in GaP is physically distinct from the strongly non-radiative dislocation cores, even though the concentrations of these two centers usually show a strong correlation in material prepared under well standardized conditions. However, it is easy to produce epitaxial GaP with τ_{mc}^{p} much lower than is indicated from the correlations with [N$_{0.75}$] and [N$_d$], showing the advent of further significant contributors to the shunt path. The data shown in Fig. 3.26 are representative of good quality material, where the influence of these additional non-radiative centers is negligible. Of the possible additional centers, only that related to V$_{Ga}$ (Sec. 3.10) has received any detailed study so far.

3.10 Vacancy-Related Recombinations

One of the most striking properties of GaP grown from stoichiometric melts SM using the liquid encapsulated Czochralski LEC technique is its very poor luminescence efficiency. Broad low energy bands predominate in the liquid *He* temperature spectra, with near gap components which are particularly weak and diffuse compared with LPE GaP. Examination of a series of crystals prepared over a range of intermediate temperatures from Ga-rich non-stoichiometric melts NSM has given further insight into this difference. The efficiency of the Zn,O-doped GaP increases monotonically with τ_{mc} as the growth temperature is reduced towards 1 100 °C, normal for solution growth SG [3.95]. (Fig. 3.28). There is also a marked increase in the density of certain saucer-shaped etch pits, S pits with increase in growth temperature T_G [3.96]. These unusual etch pits are probably associated with precipitates. They are to be contrasted with the trigonal-shaped D-pits which reveal the presence of dislocations observed in both LEC and SG GaP. The behavior of LEC GaP is more complex than SG or VPE material in a number of ways. For example, complicated annealing effects involving the key luminescence activators N [3.97] and O [3.98] have been observed. The possibility that O can exist in

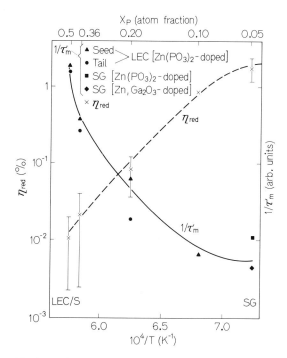

Fig. 3.28. The red photoluminescence efficiency η_{red} and transition rate through the non-radiative shunt path $1/\tau_m$ as functions of the temperature of the melt from which the GaP:Zn,O crystals were grown. Typical results for crystals grown from a stoichiometric melt by the liquid encapsulated Czochralski technique are shown on the left. The remaining LEC data were obtained from crystals grown at progressively lower temperatures from successively Ga-enriched melts and project to results typical for growth from Ga solution as shown at the right [3.95]

forms other than the well-documented O_P and associates $Zn_{Ga} - O_P$, $Li_I - Li_{Ga} -$ O_P [3.3] even in SG GaP has been shown from Li in-diffusion [3.99]. Additional O-related defects may be responsible for the inverse correlation between the O_P concentration near the junction in GaP:Zn,O and LED efficiency [3.100], given that O_P is not a dominant recombination center [3.48]. These properties remain incompletely understood.

Of particular interest is the variation of bulk material performance with T_G (Fig. 3.28) in a manner which suggests that the luminescence killer which controls τ_{mc} may involve simple microscopic defects. The influence of these microscopic defects on τ_{mc} in LEC GaP is greater than the S- or D-pits or the microscopic defect related to the incidence of D-pits discussed for LPE and VPE GaP in Section 3.8, even though the concentration of these extended defects is large in typical LEC GaP [3.101]. *Jordan* et al. [3.102] have implicated isolated V_{Ga} from analysis of the GaP solidus (Fig. 3.29), determined from coulometric titration of Ga in a series of NSM crystals. An explicit relationship has been derived between $[V_{Ga}]$ and τ_{mc}^{-1} (Fig. 3.30). These data

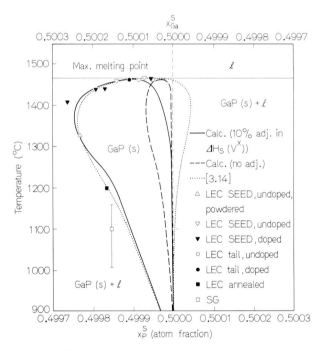

Fig. 3.29. The solidus boundary of GaP fitted to the data points obtained from coulometric titration of Ga in the indicated series of liquid encapsulated Czochralski and Ga solution grown crystals. The fits were obtained under a variety of hypotheses; dotted by assuming the presence of neutral V_{Ga} alone, dashed by a theory which considers vacancies and antisite defects in various ionization states and full line as for the dashed curve but with a 10% adjustment to the calculated virtual enthalpies of neutral vacancies in accord with experience for other materials [3.103]

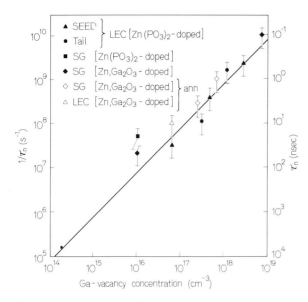

Fig. 3.30. The reciprocal of lifetime for electron capture by luminescence centers as a function of $[V_{Ga}]$ for a variety of GaP:Zn,O crystals grown as indicated over a wide range of temperatures (see Fig. 3.28) or subjected to annealing at different temperatures after solution growth. The crystals are all considered to be quenched from the growth temperature with appropriately frozen-in defect distributions, a hypothesis which has been challenged in [3.103] (see [3.102])

give an electron capture cross section attributed to V_{Ga} of $\sim 5 \times 10^{-17}$ cm^2 in LEC GaP at 300 K and also indicate that τ^e_{mc} should be in excess of 1000 μs for SG GaP prepared at 900 °C. The measured values are $\leqslant 200$ μs, although much longer values are found for τ^h_{mc} (Fig. 3.26). The comparison indicates the unimportance of this microscopic defect in epitaxial device material. This is a good illustration of the non-universality of the dominant killer centers even for a single semiconductor when prepared by very different techniques (Sec. 3.9). However, the 0.75 eV hole trap shown in Section 3.8 to be very important in n-type LPE and VPE GaP has been recently observed at high concentration in n-type LEC GaP, though it does not dominate τ^p_{mc} in this material [3.82a].

The identification of the dominant shunt path center in LEC GaP with isolated $[V_{Ga}]$ has been questioned recently. *Van Vechten* [3.103] noted that it is thermodynamically unlikely that isolated V_{Ga} exists at the high concentrations given in Fig. 3.30. If it did, there should be a large effect on the electrical conductivity, since V_{Ga} is unlikely to remain neutral as the Fermi level is varied between wide limits by changes in the conventional electrically active dopants. The center must have a significant cross section for interaction with both types of carrier if it is to control the shunt path. It is not certain whether such compensation effects occur, despite some circumstantial evidence [3.104].

This dichotomy may be resolved if we consider V_{Ga} to form an electrically neutral associate, perhaps with a native donor. *Van Vechten* [3.103] estimated the enthalpy of formation of such native defect complexes in GaP. He concluded that most of the vacancies remaining at 300 K should be tied up in neutral complexes which become thermodynamically stable during cool-down from the growth temperature. Few vacancies should escape to the outer surface before they can complex with charge-conjugate defects. Such complexes form preferentially because of the long-range Coulomb attraction between the constituents, just as for the well-known isoelectronic associates Zn_{Ga}-O_P and Li_I-Li_{Ga}-O_P [3.3]. These latter associates are efficient luminescence activators. Presumably, the dominant associates involving native defects in LEC GaP are strongly non-radiative because of the much greater importance of the MPE process (Sec. 3.7) for them. We have noted in Section 3.8 that such associates may form an important class of prototype non-radiative recombination centers. The vacancy-related centers recently studied by *Chiang* and *Pearson* [3.104] in GaAs are totally non-radiative. From the above discussion, it is not clear whether these centers are isolated vacancies as claimed or consist of associates involving vacancies. Associates containing vacancies are particularly susceptible to alteration by in-diffusion of appropriate species, like Li or Cu. It may be that many of the broad low energy luminescence bands in GaAs at present attributed to associates containing vacancies involve instead similar associates after capture of such mobile contaminant species. Evidence for an involvement of Cu in associates previously attributed to V_{As} in GaAs was obtained very recently [3.105]. We have in mind the possibility that radiative complex centers in covalently bonded semiconductors may generally consist of neutral associates of substitutional species, like the well-known activator $Zn_{Ga}-O_P$ in GaP. There is now a great deal of evidence for preferential association in many pairs of point defects which possess attractive long range interactions, even interstitial host atoms according to suggestions from recent infrared spectra in GaP [3.106]. Direct evidence was found recently for significant pairing of V_{Ga} with the Si_{Ga} donor in GaAs and GaP [3.107]. It is inferred that not all the V_{Ga} are so paired. However, the remainder could be present in associates which are not readily detected through local mode infrared absorption. Vacancies can migrate to donors at only 150 °C in ZnSe [3.108] and it is possible that annealing observed for local modes and deep electrical traps in GaP near 250 °C may have a similar interpretation [3.107].

Unfortunately, it is very difficult to obtain proof for these general ideas. As far as LEC GaP is concerned, *Van Vechten* [3.103] has noted that there may be many anti-structure defects P_{Ga}^{2+} and Ga_P^{2-}, which can behave as donors and acceptors with approximately effective-mass-like first ionization energies. He favors the extended neutral $\langle 110 \rangle$-axis defect $V_{Ga}^- P_{Ga}^{2+} V_{Ga}^-$ as the dominant shunt path center responsible for the correlation in Fig. 3.30 and notes that small defects with this symmetry have been seen in TEM studies in GaAs [3.109]. Although the calculated solidus can be fitted to the experimental data in Fig. 3.29 using either the isolated V_{Ga} or the native defect associate model,

Van Vechten considers the derived entropy for Schottky defects to be much greater than expected for neutral vacancies. The full curve in Fig. 3.29 is an estimate from the microscopic cavity model of *Van Vechten* [3.103]. The calculated virtual enthalpies of the neutral vacancy have been reduced by 10% in accord with an estimate of the accuracy of this model and some empirical data on Si, Ge and CdS. This model must be regarded as speculative, since the TEM evidence for this defect is not conclusive. There are difficulties over the constancy of the lattice parameter between GaAs prepared by different procedures for any model which predicts densities of defects containing V_{Ga} covering the full range in Fig. 3.30.

3.11 Dark Line Degradation in Injection Lasers

The injection lasers of greatest practical importance contain a double heterostructure DH. The p-type active region of GaAs is made very thin, perhaps $\leqslant 1$ μm sandwiched between two adjacent layers of $Al_xGa_{1-x}As$ (Fig. 3.31). If $x \sim 0.3$, there are steps of ~ 0.25 eV and 0.05 eV in the conduction and valence bands which serve to make electron injection predominate under forward bias

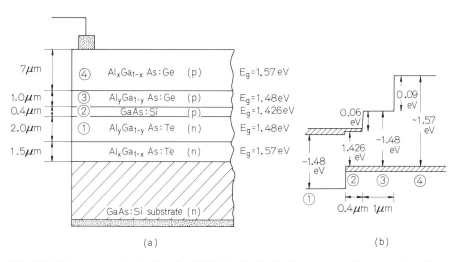

Fig. 3.31. (a) A cross-sectional view of an (AlGa)As-GaAs double heterojunction laser with a thin GaAs active layer, in which most of the injected electrons and holes are confined, sandwiched between wider regions of $Al_yGa_{1-y}As$ with the wider band gaps shown. This special structure with two compositions of (AlGa)As was designed to show that thermal activation of electrons injected through the *pn* junction between regions (1) and (2) occurs over the 0.06 eV conduction band step between regions (2) and (3) shown in part (b). The high current density, good heat sinking and freedom from threading dislocations necessary for long-lived continuous operation with 300 K heat sinks are obtained with a geometry in which the current is confined below a narrow stripe electrode with the optical facets of the laser cleaved perpendicular to the stripe axis [3.182]

and confine the minority electron population to this narrow layer, building up a very strong population inversion. The emitted light is also confined to the active layer by a dielectric waveguide effect, thereby enhancing stimulated emission. These features, combined with the heat sinking advantages of a narrow stripe geometry of contact electrodes, have allowed the construction of injection lasers which can be operated cw with their heat sinks at 300 K [3.137], an important technological development. Such behavior is possible only because the $GaAs - Al_xGa_{1-x}As$ heteroboundary exhibits an untypically low recombination rate due to interface states, below 10^4 cm s^{-1} according to a recent estimate [3.110], so that minority carrier injection is almost as efficient as in a homojunction diode. A major contributing factor to the near ideal junction behavior is the unusually close lattice match across this heteroboundary. The match is essentially perfect at crystal growth temperatures near 800 °C, while the GaAs has the smaller lattice constant at 20 °C, with a discrepancy of $5 \times 10^{-2}\%$ for $x = 0.36$. Lasers with pure GaAs active layers operate in the near infrared, at about 9000 A. Use of $Al_{0.36}Ga_{0.64}As$ active layers and $Al_{0.8}Ga_{0.2}As$ confinement layers has allowed operation to the shortest wavelength yet achieved, 7610 A for cw operation with 300 K heatsink [3.111] a wavelength within the normally quoted eye sensitivity range. As expected for direct gap recombinations, the external power efficiencies are much higher than Table 3.1, $\geqslant 20\%$ for both optically and electrically pumped laser action [3.112]. Weak heteroboundary recombination makes the DH structure well suited to the study of recombination mechanisms in GaAs. The form of the PL spectrum can be related to absorption through detailed balance [3.110]. Coulomb interactions produce a significant narrowing of both EL and PL spectra [3.113] even at 300 K in DH structures containing exceptionally lightly doped active layers, $N_D - N_A \sim 10^{15}$ cm^{-3}.

Hopes for immediate applications of these DH injection lasers were frustrated by the discovery of a very rapid degradation mechanism which was linked to the appearance of dark lines DL in luminescence topographs [3.114]. Transmission electron microscope studies [3.115, 116] have shown that these dark lines involve complex networks of dislocations which are initiated from defects originally present in the layer [3.117] (Fig. 3.32). These defects may be threading dislocations, interfacial defects arising for example from surface oxidation occurring between the intervals of LPE layer growth in the usual multiple Ga solution sliding boat system, or defects at the cleaved crystal facets which form the mirrors of the Fabry-Perot laser cavity [3.118]. These surface defects may arises from damage under the intense optical energy density in the laser cavity [3.119]. The latter effect may be minimized and the full benefit of extreme carrier confinement in very narrow active layers down to ~ 0.1 μm realized by the use of a more complex 5-layer structure involving a large optical cavity LOC [3.120]. The incidence of the rapid DL degeneration in the bulk may be reduced by scrupulous procedures which minimize interfacial contamination during growth [3.121], by careful wafer handling and device fabrication procedures designed to minimize accidental strains which may lead

to dislocation generation and, most important, by the selection of GaAs substrates with very low dislocation densities as the basis for the DH laser technology. Optical pumping experiments have shown that it is not necessary to
have a *pn* junction for DL development, though minority carrier generation
is necessary [3.122]. The DL generation may be inhibited by the preparation
of DH devices with low built-in strain at 20 °C by the use of minimal values
of x in the ternary alloy, perhaps $x \leqslant 0.2$, or a quaternary alloy including
$\sim 1.8 \times 10^{-5}$ atom fraction P which gives zero interfacial strain [3.123] essentially independent of [Al]. It has been estimated that such DH lasers may have

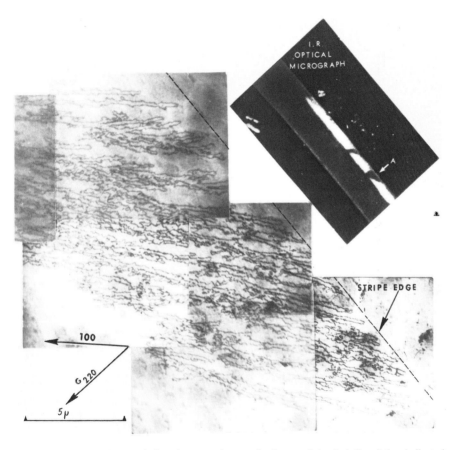

Fig. 3.32. Bright field transmission electron micrograph of part of the dark line defect indicated
by A on the infrared luminescence micrograph of an (AlGa)As-GaAs double heterojunction laser
shown in the upper right corner. The defect is confined to the laser stripe area containing the injection
current. It consists of a high density ($10^8 - 10^9$ cm^{-2}) network of dislocation dipoles with their
long axes parallel to $\langle 100 \rangle$ or $\langle 210 \rangle$ directions. The Burgers vectors are $a/2 \langle 011 \rangle$ type inclined at
45° to the $\langle 100 \rangle$ junction plane. No evidence of precipitate decoration has been seen. Defect
propagation is attributed to rapid non-conservative dislocation climb [3.115]

very long lifetimes $> 10^5$ hrs when operated with *junction* temperatures of only 26 °C [3.124]. However such estimates are necessarily based on extra-polations of accelerated life tests which may be open to doubt because of the difficulty in establishing a simple universal form for the temperature dependence of the degradation rate containing a single exponential factor [3.125].

The mechanism by which the DL defects propagate from the initiating source, most frequently a threading dislocation, is a matter of great topical interest. Initial TEM studies established the DL defects as massive networks of dislo-cation dipoles usually with $\langle 100 \rangle$ orientation (Fig. 3.32) [3.115]. The Burgers vectors are $a/2 \langle 110 \rangle$, as is normal in the zincblende structure. The unusual network propagation direction has been attributed to multiplication by a non-conservative climb process in which pure edge dislocations consume vacancies [3.115]. Other workers claim that the dislocation dipoles and loops which lie outside them are predominantly of interstitial (extrinsic) type [3.116]. In either case, there are considerable difficulties with the notion that the main factor in DL defect propagation is dislocation climb [3.126]. The very rapid propagation, which may proceed in fits and starts depending upon the laser drive and ambient light level, is hard to explain. It would seem to require that vacancies or inter-stitials must preferentially swarm to the tip of the dipoles, which should impose a low limit on the propagation velocity set by the diffusivity of these microscopic defects. The climb model also implies the heteroboundaries to be rich sources of vacancies and interstitials. This seems unlikely both on the grounds of their electronic inactivity [3.110] and from the observation that inclined dislocations which jog to relieve the interfacial strain invariably do so within a single plane, the glide plane of the edge dislocation [3.90]. *Matsui* et al. [3.126] recently proposed a model for DL defect propagation in which screw dislocation segments are formed by cross-slipping on alternate $\{111\}$ planes to give the observed overall $\langle 100 \rangle$ propagation direction (Fig. 3.33). This conservative process may be very fast. Various kinematic possibilities exist to explain other DL orientations sometimes observed. Dipoles formed mainly by this glide process can thicken and twist by a combination of further glide and climb of the pairs of screw segments which made up the zig-zag structure of the original dipole. Various mechanisms exist by which the loops observed within and with-out the main dipoles can be formed during this stage, producing a most complex overall structure (Fig. 3.32). Point defects can interact along the whole length of the dipole during this thickening process, so there is no significant limitation from diffusion processes.

The DL defects are known to be strong non-radiative sinks of minority carrier recombination, producing local unpumpable regions which absorb the coherent light generated in defect-free regions [3.127, 128]. The sinks are a few μm diameter and the excess recombination current through these regions can cause local heating of order 100–200 °C. Temperature rises of this order are suggested both from measurements with a thermal plotter and from model calculations [3.129]. The heat is generated essentially by the MPE process described in Section 3.7. There are indications from preliminary measurements on GaP

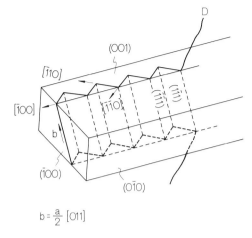

$$b = \frac{a}{2} [011]$$

Fig. 3.33. A schematic representation of a conservative gliding screw dislocation propagation model for the ⟨100⟩-oriented dark line defects characteristic of rapid performance degradation in (AlGa)As-GaAs double heterojunction lasers (Fig. 3.32). The inclined initiating dislocation D is mainly screw type near the active layer. This section can glide on alternate {111} planes to create a ⟨100⟩-oriented dipole of mixed dislocations. The dipoles can elongate rapidly by this conservative mechanism, then thicken and twist by further glide and climb to produce the typically complex forms shown in Fig. 3.32 [3.129]

[3.130] that this mechanism is an even more serious problem for wider gap semiconductors, as might be expected [3.78] since the energy to be dissipated, $\sim E_g$, increases faster than the critical stress for dislocation motion or the thermal diffusivity. Local heating at the DL defects produces strong thermal stress gradients which may well provide the driving force for the cross-slip which propagates them and is the main feature of the performance degradation on this model. The built-in stress due to heteroepitaxy in the $GaAs - Al_xGa_{1-x}As$ is evidently significantly below the critical stress for dislocation generation in this system. The model calculations [3.129] indicate thermal stresses of $\sim 10^9$ dyne cm^{-2} under typical excitation densities of 10^4W cm^{-2}, about an order greater than the heteroepitaxial interfacial stress [3.131]. Thus, the existence of degradation-resistant regions is simply explained by the absence of threading dislocations or other defects where heating from local non-radiative recombination may augment the built-in interfacial stress above the critical value, estimated to be below 10^9 dynes cm^{-2} at the relevant temperatures [3.129]. The critical role of local recombination explains the observed localization of the DL defect development to that portion of the threading dislocation under the stripe contact and mainly within the active region. It also explains the extreme sensitivity of the propagation rate to the injection level of the laser.

The recent rapid developments in our understanding of these complex phenomena have led to substantial improvements in the operating life of room temperature cw injection lasers, which have already been reproduced in several laboratories in different countries. We can now claim the GaAs DH laser as a thoroughly practical light source [3.132]. Such a source is currently needed for the rapidly emergent new technology of optical communications, in which information is transmitted along glass fibers in the form of a guided beam of near infrared light.

3.12 Optical Effects of Two-Dimensional Confinement of Electrons and Holes in Semiconductors

The DH injection laser has recently drawn attention to the behavior of electrons and holes confined within very narrow slices of a semiconductor. Such systems can exhibit striking characteristic optical phenomena. Before reviewing these, we will briefly consider complementary evidence obtained from electrical transport measurements of the quantum behavior of electrons confined within narrow one-dimensional potential wells. Although these data are obtained from structures related to an entirely different type of device in a different semiconductor Si, it is important to recognize the close relationship with the optical properties measured in GaAs. The quantum effects of confinement become noticeable when the well width a becomes small enough so that $ka \lesssim 1$, where k is the wave vector of the electron. Striking effects can be observed in certain electrical properties, for example in the narrow surface inversion layer of insulated gate field effect transistors in Si [3.133].

Two-dimensional localization of the electrons in the triangular one-dimensional potential well produced by the electric field normal to the surface produces two ladders of allowed energies in n-type Si associated with the two possible values of electron effective mass parallel to the surface. Thus, with a $\langle 111 \rangle$ surface

$$E(k) = E_i + \frac{\hbar^2 k_t^2}{m_t} + \frac{\hbar^2 k_2^2}{m_2} \tag{3.9}$$

where $m_2 = \frac{1}{3}(m_t + 2m_l)$, E_i is the energy normal to the surface and m_t, m_l are the usual ellipsoidal conduction band masses in Si. Application of a magnetic field parallel to the surface splits each sub-band into Landau levels. Increase in gate voltage raises the carrier concentration in the inversion layer, causing the Fermi level to pass through successive Landau levels. The conductance falls to a minimum when each Landau level becomes completely filled. The amplitude range of the resulting de Haas-Shubnikov oscillations in conductance can exceed 10^6 and constitute the largest quantum magnetoresistance effect yet known. In addition, structures are observed in the dependence of field effect mobility on gate voltage at zero magnetic field. These are related to two effects, both dependent on traps at the Si/SiO$_2$ interface. First, fluctuations of the charge state of these traps give a random potential and hence a mobility edge separating localized from extended states in the two-dimensional conduction band [3.134]. The system shows an Anderson transition as the Fermi level moves through this mobility edge in response to the gate voltage. The second, "electric freezeout" effect is attributed to the increased effectiveness of localization in discrete surface traps about 10 meV below the band edge as the Fermi level increases with bias [3.135].

Although these electrical effects in Si inversion layers are very striking, it has been found that the quantum mechanical properties of electrons in such structures can be explored in even greater detail by optical techniques, as is frequently true. The precise control of epitaxial growth on the scale of a few atomic layers afforded by molecular beam epitaxy (MBE) has enabled the growth of essentially perfectly flat layers of GaAs sandwiched between layers of $Al_xGa_{1-x}As$ which have much wider bandgap. It is possible to grow successively scores of GaAs layers of constant thickness, say ~ 100 Å, interleaved with $Al_xGa_{1-x}As$ layers which may be made thick enough to electronically isolate adjacent GaAs layers or not, just as desired. The total thickness of the set of GaAs layers is then sufficient for examination of the GaAs absorption edge [3.136]. The thicker layers show the classical exciton line and continuum familiar for a direct transition in bulk material (Fig. 3.34). As the layer thickness is reduced, the edge shifts slightly to higher energies and well-defined

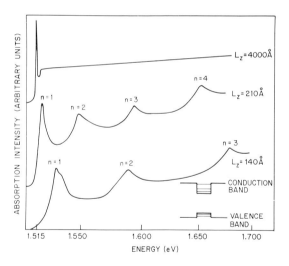

Fig. 3.34. The absorption edge of thin layers of GaAs, thickness L_z, interleaved with $Al_xGa_{1-x}As$ layers sufficiently thick to isolate the electron and hole bound states formed in the narrow one-dimensional conduction and valence band wells localized at the GaAs layers. The extra peaks present for the thinner layers arise from exciton transitions which are allowed only between bound states of the same quantum number. The structure in the lowest energy peak for $L_z = 140$ Å is the only evidence here of a second localized exciton series involving heavy holes. Sandwiches containing up to ~ 50 GaAs layers of constant thickness are necessary to obtain such well-defined absorption with the smallest values of L_z [3.136]

peaks appear extending considerably above the absorption edge. These are exciton peaks associated with the ladders of allowed electron and hole states in the one-dimensional potential well. These states are qualitatively the same as those described above, but now the potential well is formed by the near ideally abrupt discontinuities in valence and conduction band energies at the metallurgical interfaces in the heterostructure. Allowed exciton transitions occur between electron and hole states with the same quantum numbers. The conduction band states are very simple in GaAs, with a single effective mass appearing in 3.9. However, two exciton series appear associated with the light and heavy holes in the valence band.

These spectra are important both for giving graphic support to the quantum-mechanical theory of electronic behavior in a situation which is very simple to treat theoretically, as well as for providing a way of estimating the apportionment of the band discontinuities at the heteroboundaries, ~ 0.85 in the conduction band in this system. This parameter is important for the design of heterostructure lasers, since we are dealing with the only system which supports cw laser action at 300 K to date [3.137]. Further exploitation of the technique is possible, for example in the study of coupled layers [3.138], resonant tunneling in a series of coupled layers [3.139] and laser action on the quantum levels [3.140]. Double heterostructure injection lasers have been made by MBE, though not so far with record performance [3.141].

3.13 Gallium Nitride

We have discussed in Section 3.1 general trends in the properties of semiconductors in a given class which predict that GaN should have an energy gap appreciably greater than III-V compounds whose atomic constituents lie lower in the periodic table. This prediction was not thoroughly verified for GaN until the late 1960s. The problem has been the difficulty of preparation of single crystals of adequate size and purity, also an effect consistent with the trends described in Section 3.1. A fairly satisfactory VPE technique has now been evolved in which N is transported as NH_3 and the Ga as a chloride species, with deposition at $\sim 900\ °C$ [3.142]. Freely grown hexagonal needles have been produced in a similar system but at much higher temperature $\sim 1100\ °C$ [3.143]. There are two main difficulties. First, the deposition is kinetically controlled, since the equilibrium vapor pressure of N_2 over GaN is very high at the growth temperature. This makes the growth rate and crystal quality particularly sensitive to the detailed parameters of the growth reactor. In particular, growth of lightly doped material is very sensitive to the geometry and temperature gradients in the vicinity of the outlet of the Ga supply tube and the substrate, which should be in close proximity [3.144]. The possibility of obtaining semi-insulating material by co-doping with acceptors such as Zn [3.145], Mg [3.146] and Be [3.147] depends critically upon these factors, which must be adjusted empirically. The electrical transport properties of undoped layers are strongly influenced by the growth rate and crystalline perfection of the layers [3.148]. The second problem is the lack of a suitable substrate for homoepitaxy. The $\langle 0001 \rangle$ surface of sapphire is used most frequently for heteroepitaxial growth, although this is not the optimum orientation [3.149]. Unfortunately, there is significant lattice mismatch between sapphire and GaN. Since the thermal coefficient of expansion of sapphire is about twice that of GaN, the layers grow under considerable strain, sometimes cracking the substrate upon cooling to 300 K after growth. The strain is correlated with an Urbach-type tail at the intrinsic absorption edge [3.150]. Despite all these difficulties, relatively large GaN layers of quality sufficient for detailed optical analysis have been prepared. Unlike the lower gap III-Vs GaN grows in the wurtzite habit, with the c axis normal to the substrate. Undoped GaN may be water-white as a 20 μm layer, invariably n-type with carrier concentration usually well into the 10^{18} cm^{-3} range whether grown by VPE [3.142] or LPE from Ga solution [3.151]. Exceptionally, undoped material has been prepared with $N_D - N_A$ in the low 10^{17} cm^{-3} range [3.144]. Semi-insulating material is invariably achieved by compensation at much higher combined impurity concentrations [3.145]. All the circumstantial evidence suggests that the general tendency of GaN to exhibit large $N_D - N_A$ is associated with native defects, most likely V_N. Certainly, the best quality GaN does not contain any dopant likely to form a donor species at a concentration comparable with the observed free electron density [3.149]. This self-compensation is also presumed responsible for the general failure to prepare p-type GaN, despite considerable efforts at low N overpressures [3.145–147]. Undoped high n-type and particularly Zn-doped compensated layers of exceptional morphological quality have been prepared recently using high NH_3 partial pressures and eliminating H_2 to minimize decomposition of the GaN immediately after synthesis [3.152].

As described thus far, GaN is a great disappointment. At the outset, most workers believed that the principal difficulty was the preparation of the material in single-crystal form suitable for assessment and device fabrication. Unfortunately, this difficulty has been mastered only to reveal a problem of self-compensation like that endemic in the more ionic AB compounds (Sec. 3.1). Some people regard GaN and the still wider gap III-V AlN as II-VI – like III-V compounds and point to their hexagonal crystal structure as evidence of an instability against intrinsic defect formation intimately connected with the self-compensation problem. However, it is more likely that the difficulty is simply that the bandgaps of these III-V nitrides are so large that the gain in energy obtained during crystal growth by the generation of extra defects which can electrically compensate deliberately added dopants is larger than the enthalpy of intrinsic defect formation. The latter does not increase as fast as the former within a given class of compounds, whichever of the directions towards a larger bandgap mentioned in Section 3.1 is followed [3.3, 153]. The two directions give very different rates of increase of bond ionicity with bandgap. It is true that several of the more ionic AB compounds tend to grow in the hexagonal rather than cubic lattice structure. However, the differences between the defect enthalpies in these lattices are relatively minor, as expected since the two crystal structures differ only in a rather subtle manner up to the configurations of third nearest neighbors. Much larger differences in defect enthalpy are exhibited between the tetrahedrally and octahedrally coordinated lattice structures, as is now well known both experimentally [3.75] and theoretically [3.154]. Further evidence of the mild consequence of the change from zincblende to wurtzite lattice as far as self-compensation is concerned may be found in from the fact that many II-VIs grow readily in both forms with equal difficulties for amphoteric doping. Other semiconductors such as SiC can also be obtained in both structures with quite wide changes in E_g but approximately equal *ease* of amphoteric doping [3.3].

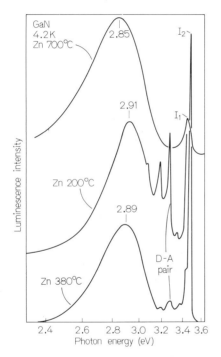

Fig. 3.35. Photoluminescence spectra from VPE GaN doped progressively more heavily with Zn according to the indicated temperatures of the Zn source. As [Zn] increases, the broad band in the blue-violet becomes progressively stronger relative to the structured ultraviolet edge luminescence. The edge luminescence is labeled in accordance with practice for II—VI semiconductors, I_2 and I_1 arising from the decay of excitons bound at neutral donors and acceptors, respectively. The broad band involves recombinations at a deep Zn-related acceptor, not simply Zn_{Ga} which is believed to contribute to the UV D-A pair luminescence. The identity of this deep acceptor is unknown but it has been speculated that it could involve the anti-structure substituent Zn_N. *Lagerstedt* and *Monemar* [3.158] have observed weak structure in a probably analogous broad luminescence band in GaN:Cd, with no-phonon line at 2.937 eV at 1.6 K [3.159]

The fact that various groups have persisted in the study of electrolumin-escence of GaN despite the lack of a technology for *pn* homojunctions is a reflection both of the absence of easy alternatives for cosmochrome displays and the encouragement obtained from many of the early results on the basic optical and phosphor properties of GaN. Early optical absorption studies [3.150] suggested that E_g is near 3.5 eV at 300 K and is direct [3.155]. This was confirmed in more detailed studies on the exceptionally lightly doped material mentioned above [3.144]. Optical reflectivity has revealed intrinsic exciton structure very similar to ZnO with $E_g = 3.50 \pm 0.01$ eV, small spin-orbit and crystal field splittings in the valence band maximum and an exciton binding energy of ~ 30 meV [3.144]. The low temperature ultraviolet edge luminescence [3.156] (Fig. 3.35) is generally similar to other III-V and II-VI direct gap semiconductors, dominated by donor and acceptor bound exciton and d–a pair recombinations (Sec. 3.2). A most surprising feature is the apparent identity of the binding energies of Be, Mg [3.157], Zn [3.156, 159] and Cd [3.158, 159] acceptors, all falling close to 200 meV. More relevant for LED applications is the efficient broad-band luminescence in the blue or blue-green observed at high concentrations of several of these acceptors and peaking near 2.85 eV for Zn (Fig. 3.35). Like many of the wider gap II-VI compounds, such appropriately doped GaN shows very good visible luminescence efficiency at 300 K under band-to-band excitation by photons or energetic electrons. The relatively slow decaying blue luminescence of GaN:Zn has been attributed to tunneling transitions from conduction band tail states in donor-rich regions to nearby acceptors ~ 0.7 eV above the valence band [3.160]. This model provides an explanation for the kinetic and spectral behavior reminiscent of classical d–a pair recombinations reported for GaN:Zn [3.161]. Several electronic properties of GaN:Zn become suddenly improved with increasing [Zn], coincident with a dramatically abrupt transition from *n*-type to semi-insulating behavior[3]. One plausible but unproved explanation [3.162] is that excess Zn enters the lattice at N vacancies, at once removing the V_N triple donor and replacing it by a triple acceptor. The consequent reversal of the sense of the internal stress and close electrical compensation may account for changes in the form of the extrinsic photoconductivity spectrum and for an inability to produce *p*-type GaN controlled by the shallow Zn_{Ga} acceptor.

The absence of *p*-type conductivity means that non-homojunction structures must be used for GaN electroluminescence (Fig. 3.36). Until recently, the most successful device work has been performed at RCA Laboratories using MIS configurations [3.163] in which the light is most efficiently generated in forward bias, originating mainly at the $i–n$ interface within the relatively thick VPE GaN [3.145]. Power efficiencies of up to 0.1% have been claimed in the blue-green [3.146, 152] and more recently up to ~ 0.3% for green and blue devices and more rarely up to 1% in the green [3.164]. No details of the special devices used in the recent work [3.152, 164] are yet available, although it is reported that they are neither classical *pn* nor MIS structures. A key feature is an

---

[3] See *Notes Added in Proof*, p. 126, Note (3).

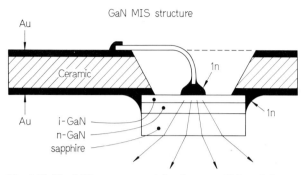

Fig. 3.36. The MIS structure used for the most efficient GaN LED. The *n*-layer may be 100 μm thick to reduce the effect of strains generated by lattice mismatch at the GaN-Al₂O₃ heteroboundary on the active region at the *i−n* interface. The highly conducting undoped *n*-layer (~10^{-3} Ωcm) is biased negative. The color of the electroluminescence can vary appreciably depending upon the layer thicknesses, the Zn doping level in the *i* layer which can give it brownish-orange color and the way in which the light is extracted from the chip [3.163]

operating voltage of only 2–4 V, about 20% of the range needed for most earlier high efficiency GaN devices [3.165]. The color of the EL can be varied between wide limits, depending on the layer thicknesses, Zn doping and growth conditions [3.165]. Very high electric fields occur at the *i−n* interface in Fig. 3.36 [3.165]. The high field from the large potential drop at the cathodic edge of the I layer is localized by sharp points and ridges found from SEM studies at both layer edges [3.166] (Fig. 3.37). The Fowler-Nordheim type forward I, V characteristic implies current control by tunneling through a triangular potential barrier at the cathode [3.167]. It has been suggested that electrons gain energy in the junction field and produce impact ionization at the deep Zn-related acceptor activators [3.167]. This gives rise to microplasma-type luminescence topograms analogous to those studied in detail in *Si* diodes much earlier [3.168].

Devices of the form shown in Fig. 3.36 with a barely Zn-compensated I layer [3.169] show power efficiencies up to 3×10^{-4}, a record value for blue light before the recent reports from the LEP group [3.152, 164]. A curious feature is that this blue EL can appear with a peak energy substantially higher than the applied (forward) bias. Various explanations have been advanced for this anti-Stokes behavior, involving some form of double impact excitation of electrons from valence band or Zn centers by hot electrons which have tunneled through the *i−n* interface barrier. These ideas are supported by an observed cubic variation of the EL intensity with diode current. The avalanche breakdown and impact ionization aspects of the model are conceptually very similar to those proposed for EL in reverse bias ZnSe:Mn Schottky barriers and investigated in detail by *Allen* [3.170]. The main difference is that the avalanche breakdown is usually relatively uniform in ZnSe:Mn and sub-grain boundaries play no essential role. Other possible applications interests in GaN are solar blind photoemitters [3.171] and negative affinity electron emitters [3.172].

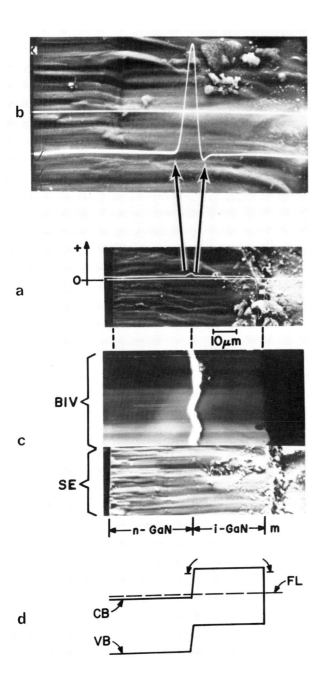

Fig. 3.37a–d. Beam induced voltage scanning electron microscope topographs in a GaN MIS structure similar to Fig. 3.36. The internal electric fields present at the $m-i$ and $i-n$ interfaces separate the charges generated by the electron beam and produce the changes in beam-induced voltage shown in (a) superimposed on the secondary emission topograph. The changes near the critical $i-n$ interface due to changes in chemical composition are shown amplified in (b), and in analog form in (c). The polarity and magnitude of these topographs suggest the band structure profile shown in (d). The cathodoluminescence efficiency is greatest near the $i-n$ interface where the material is just compensated and Zn concentration quenching is insignificant [3.166]

The performance of these GaN MIS lamps is not yet competitive with the yellow, orange and red LEDs devices listed in Table 3.1 although the most recent reports from the LEP group give extremely promising laboratory data [3.164]. The II-VI variants such as yellow ZnSe:Mn [3.170] have also yielded promising laboratory data, yet are regarded as uncompetitive despite the advantages of easier crystal growth of such II-VI compounds compared with GaN:Zn. However, as we have seen in Section 3.1, there is no competition from the true homojunction LEDs in the blue or blue-green. The GaN:Zn device provides one of the best direct emitters currently available for blue EL. The anti-Stokes type high current density blue emitter is suitable for small indicators and can complement the longer wavelength devices of Table 3.1 in performance if not production cost [3.173]. Another possibility is the so-called up-converter, in which three infrared quanta excite Tm^{3+} in a rare earth phosphor host to emit a single blue quantum [3.3]. Neither of these approaches seems to have very attractive commercial prospects on present day performance and degree of control, although it is too early to make definite statements about the LEP device.

A better choice may be the ZnS or $ZnS_{1-x}Se_x$ forward bias Schottky diode, which emits blue light from d − a pair recombination attributed to the $V_{Zn} - Al_{Zn}$ associate acceptor and Al_{Zn} donor [3.174]. There has been very recent improvement in our understanding and control of the conditions under which forward bias EL in such systems may have interestingly high efficiency [3.175]. Power efficiencies of $\sim 2 \times 10^{-4}$ have been obtained at useful output levels in ZnS diodes treated so that the I, V characteristic of the Schottky contact is degraded by a thin insulating barrier of controlled type beneath the Schottky electrode. This is comparable with the present record performance for blue electroluminescence from GaN MIS diodes [3.169] although it is only $\sim 10\%$ of the very recent LEP result [3.164]. However, the fabrication technology of II-IV lamps currently seems to hold the greatest promise for economic manufacture on a large scale, either in the single-crystal form considered here or in thin film or powder EL cells [3.176, 177].

3.14 Future Prospects

This review has concentrated mainly on the III-V materials from which the commercially available LEDs are currently made. The incoherent LEDs are all homojunction devices. The power efficiency of these devices, though useful, has remained low by comparison with the theoretical limit for many years now. Recent attempts to improve our understanding of this limitation have been emphasised in this review. The general nature of the problem is clear. The efficiency of near gap electroluminescence is pinned near 0.1% in practical LEDs by the limitation on the luminescence efficiency in the bulk material resulting from the strong competition of minority carrier recombinations through defects with large trap depths. These defects are present inadvertently and remain unidentified. The competition is so severe for the indirect gap $GaAs_{1-x}P_x$ or GaP LEDs used for orange, yellow or green light that the minority carrier lifetime is dominated by this totally non-radiative so-called shunt path process rather than by the luminescence activators. Considerably higher internal efficiencies of bulk luminescence are only possible either by using $GaAs_{1-x}P_x$ with $x \lesssim 0.4$, to take advantage of the rapid, direct type radiative recombination process, or by using a luminescence activator with a minority carrier trapping level sufficiently deep that thermal re-emission is slow and the activator can compete with the deep non-radiative shunt path traps on more equal terms. Unfortunately, both of these approaches can only produce red light in the technologically amenable semiconductors. The luminous equivalent of this red light is lower by an amount just sufficient to annul the improvement resulting from the increased bulk luminescence quantum efficiency. Physiological factors also favor LEDs which emit light near the center of the visible spectrum over red and, particularly blue emitters. Limitations of visual acuity occurring near the extremes of the spectral response of the human eye are particularly relevant for complex displays rather than simple indicators.

The future prospects for improvement of LED performances are therefore closely identified with the extent to which the shunt path centers can be eliminated or reduced. We have seen that it is very difficult to identify such centers. However, some recently developed techniques make it much more convenient to diagnose the presence and concentration of non-radiative recombination centers with large trap depths. Such detailed diagnoses of the dominant luminescence killer centers are now needed since sufficient time has elapsed for LED performance to have been improved as far as is likely without this information.

The alternative to the reduction of the shunt path is the use of activators with improved radiative recombination properties at a given spectral wavelength. However, the development of the subject has reached a stage where it is unlikely that this will be possible for the semiconductors in Table 3.1. Attention then turns to the development possibilities of the materials which can be used only for non-homojunction LED systems, not discussed in detail in this review. In addition to the devices which can produce blue light discussed briefly in Section 3.13, there are a number of systems which at present operate best at longer luminescence wavelengths. Of these, the recent rapid developments of performance of AC [3.176] and DC [3.177] ZnS:Mn EL devices of the thin film or powder cell type merit special consideration. These devices produce very attractive yellow-orange light, now with commercially promising brightness and life. Thus far, it has appeared that such systems will find most appropriate application in complex displays, particularly those of large surface area. For the smaller displays, where LEDs are most suitable, it is now certain that significant further improvements in basic device performance are demanded in many applications. Most of these involve increased efficiency, particularly in battery-powered systems. The wristwatch is an extreme example, where an

LED efficiency improvement of 3−5 times would be extremely advantageous. The demand for such improvements is particularly pressing because of recent successes in the use of passive displays for this and similar applications, for example in the hand-held electronic calculator.

It is imprudent to predict the future for LED improvements in quantitative terms, particularly when we must admit to so little knowledge about the real nature of the centers which limit performance. However, the 3−5-fold improvement in the average efficiency of the near-gap indirect emitters just mentioned would not appear to be implausible in commercial practice, in view of Table 3.1. Such improvements may well result from greater control of the shunt path consequent upon the deep level diagnostic work now in progress. This review emphasizes that most of the preliminary diagnostic work has been performed with GaP, although $GaAs_{1-x}P_x$ is more important commercially (Table 3.1). This is understandable in view of the state of our knowledge of the general properties of binary III-V compounds compared with the alloys. However, it is now time for the application of the deep level diagnostic techniques to the alloys, especially $GaAs_{1-x}P_x$, where we have seen that there is particularly great scope for performance improvement through better control of the shunt path recombinations.

Acknowledgements. The author is grateful to *R.Z.Bachrach*, *A.S.Barker*, Jr., *R.N.Bhargava*, *G.R.Booker*, *C.H.Henry*, *N.Holonyak*, Jr., *D.T.J.Hurle*, *R.F.Kirkman*, *H.Kressel*, *D.V.Lang*, *Y.Nannichi*, *J.Pankove*, *M.D.Sturge*, *A.M.White* and *D.R.Wight* for the provision of illustrative material, sometimes in preprint form or for discussions of various technical points.

Notes Added in Proof

Since this article was first prepared, in early 1976, there have been a few developments of key significance to certain of the properties we have discussed.

(1) The Binding Energies of Acceptors and the Free Exciton in GaP

It has been suggested in Section 3.2 that the energy gap and therefore the free exciton binding energy E_x should be increased by 11 meV compared with *Dean* and *Thomas* [3.11d]. Some 8 meV of this increase stems from an upward revision of the acceptor binding energy E_A suggested from [3.11c]. However, *Street* and *Senske* [3.187] have reported much more detailed measurements of inter-bound state transitions of acceptors from the excitation spectra of D-A pair luminescence. Their data are consistent with transitions observed in electronic Raman scattering [3.188]. Assignments made in terms of recent theoretical studies of acceptor states [3.189] suggest that no revision of E_A is needed. Thus, the increases in E_g and E_x are due to the changes in E_D alone, and are probably $\sim +4$ meV [3.11a] compared with *Dean* and *Thomas*.

(2) The Behaviour of N in $GaAs_{1-x}P_x$ (Sec. 3.4)

Considerable attention has been devoted to the dependence on the compositional parameter x of the near-gap electronic properties of $GaAs_{1-x}P_x$, both intrinsic and related to the isoelectronic substituent N. The free exciton structure is remarkably little broadened by alloy disorder, and optical absorption has confirmed the compositional dependence of the direct gap, lowest below $x=0.45$ (77 K) [3.184]. However, early indications [3.40] that electronic states whose binding is determined by short range forces are significantly broadened in the alloy have been confirmed by recent optical absorption and related measurements [3.185]. The assignments in Fig. 3.15 have been appreciably modified. The N-related state labelled NN in Fig. 3.15 is now recognized as N_X, strongly broadened by increased phonon coupling as the binding energy of the X conduction band-related exciton increases sharply with decrease in x. The experimental points just below and to the left of the label A are now associated with a second bound exciton state of isolated N, derived from the Γ conduction band minimum. This state can be followed down to $x \sim 0.35$ but is probably unbound for lower x. The NN states must exist below $x \sim 0.85$. However, they are so heavily broadened and the luminescence from them is weakened by decreasing tunnelling from the much more tightly bound N states so that no optical effects of them are now recognized at these lower values of x. The presence of the additional N_Γ state is now interpreted in terms of a two component binding potential [3.186].

(3) Ion Implantation in GaN

Comparison of the optical properties of GaN implanted by 35 different elements [3.190] confirms that Zn provides the most efficient recombination center (Sec. 3.13). A band near 2.15 eV (78 K) is attributed to the byproducts of radiation damage remaining after the anneal at 1050 °C in flowing NH_3. Hole injection efficiencies of 50% have been reported for GaN LEDs [3.191].

References

3.1   H. Welker: Z. Naturf. **11**, 744 (1952)

3.2   J. J. Hopfield: Phys. Chem. Solids **15**, 97 (1960)

3.3   A. A. Bergh, P. J. Dean: *Light Emitting Diodes* (Clarendon Press Oxford 1976)

3.4   M. G. Craford: *Progress in Solid State Chemistry*, Vol. 8 ed. by J. O. McCaldin, G. Somorjai (Pergamon, Oxford, 1973) p. 127

3.5   S. S. Devlin: *Physics and Chemistry of II-VI Compounds*, ed. by M. Aven, J. S. Prener (Wiley, New York 1967) Chap. 11

3.6   Direct evidence that the shallow cation substitutional acceptors Li and Na also form exceptionally shallow donors, presumed interstitial, in CdS is presented in [3.84]

3.7   Structured d–a pair spectra were first identified in GaP; D. G. Thomas, M. Gershenzon, F. A. Trumbore: Phys. Rev. **133**, A269 (1964)

3.8   The term edge luminescence was first introduced to describe the series of narrow bands in certain II-VI semiconductors which we now know to be due to electron-hole recombinations at distant pairs of shallow donors and acceptors and of free electrons at shallow neutral acceptors with replication by LO phonons, as first suggested by L. S. Pedrotti, D. C. Reynolds: Phys. Rev. **120**, 1664 (1960)
       It is now frequently used to include the narrow line luminescence found still closer to the band gap due to the recombination of free and bound excitons (Fig. 3.4)

3.9   A. Onton: Phys. Rev. **186**, 786 (1969); A. Onton, R. C. Taylor: Phys. Rev. B**1**, 2587 (1970)

3.10   H. C. Casey, Jr., F. Ermanis, K. B. Wolfstirn: J. Appl. Phys. **40**, 2945 (1969)

3.11   P. J. Dean, D. C. Herbert: J. Luminesc. **14**, 55 (1976)

3.11a A. C. Carter, P. J. Dean, M. S. Skolnick, R. A. Stradling: J. Phys. C. (to be published)

3.11b W. Scott, J. R. Onffroy: Phys. Rev. B**13**, 1664 (1976)

3.11c W. Berndt, A. A. Kopylov, A. N. Pikhtin: JETP Lett. **22**, 578 (1975)

3.11d P. J. Dean, D. G. Thomas: Phys. Rev. **150**, 690 (1966)

3.12   P. J. Dean: *Progress in Solid State Chemistry*, Vol. 8, ed. by J. O. McCaldin, G. Somorjai (Pergamon, Oxford 1973) p. 1

3.13   D. J. Ashen, P. J. Dean, P. D. Greene, D. T. J. Hurle, J. B. Mullin, A. M. White: J. Phys. Chem. Solids **36**, 1041 (1975)

3.14   A. M. White, P. J. Dean, K. M. Fairhurst, W. Bardsley, E. W. Williams, B. Day: Sol. State Commun. **11**, 1099 (1972)

3.14a J. W. Allen, P. J. Dean, A. M. White: J. Phys. C.: Solid State Phys. **9**, L113 (1976)

128   *P. J. Dean*

3.15 W. Schairer, E. Grobe: Sol. State Commun. **8**, 2017 (1970); also
      A. M. White, I. Hinchliffe, P. J. Dean, P. D. Greene: Sol. State Commun. **10**, 497 (1972)
3.16 W. Schairer, M. Schmidt: Phys. Rev. B **10**, 2535 (1974)
3.17 F. Williams, D. Bimberg, M. Blatte: Phys. Rev. B **7**, 2473 (1973);
      M. Illegems, R. Dingle, L. W. Rupp, Jr., J. Appl. Phys. **46**, 3059 (1975)
3.18 Sh. M. Kogan, B. I. Sedunov: Fiz. Tverd Tela **8**, 2382 (1966) (Engl. Transl. Sov. Phys. – Solid State **8**, 1898 (1967)
3.19 T. N. Morgan: Phys. Rev. Lett. **21**, 819 (1968)
3.20 P. J. Dean, D. D. Manchon, Jr., J. J. Hopfield: Phys. Rev. Lett. **25**, 1027 (1970)
3.21 A. S. Barker, Jr.: Phys. Rev. B **7**, 2507 (1973)
3.21a R. C. Newmann: Advan. Phys. **18**, 545 (1969); W. G. Spitzer: Festkörperprobleme Bd. XI/1, 1971
3.22 P. J. Dean, R. A. Faulkner, S. Kimura: Phys. Rev. B **4**, 1926 (1971)
3.23 S. F. Ross, M. Jaros: J. Phys. C: Sol. State Phys. **1**, L235 (1974)
3.24 R. F. Leheny, J. Shah: Phys. Rev. B **12**, 3268 (1975)
3.25 P. J. Dean: J. Luminesc. **1**,**2**, 398 (1970)
3.26 J. C. Campbell, N. Holonyak, Jr., M. G. Craford, D. L. Keune: J. Appl. Phys. **45**, 4543 (1974)
3.27 R. N. Bhargava: IEEE Trans. ED-**22**, 691 (1975)
3.28 F. P. J. Kuijpers, L. Blok, A. T. Vink: J. Cryst. Growth **31**, 165 (1975)
3.29 R. A. Logan, P. J. Dean, H. G. White, W. Wiegmann: J. Appl. Phys. **42**, 2328 (1971)
3.30 D. I. Kennedy: Private communication (1975)
3.31 E. Cohen, M. D. Sturge, N. O. Lipari, M. Altarelli, A. Baldereschi: Phys. Rev. Lett. **35**, 1591 (1975)
3.32 J. J. Hopfield, D. G. Thomas, R. T. Lynch: Phys. Rev. Lett. **17**, 857 (1965)
3.33 J. L. Merz, R. A. Faulkner, P. J. Dean: Phys. Rev. **188**, 1228 (1969)
3.34 P. J. Dean: Unpublished data
3.35 D. Bimberg, P. J. Dean, F. Mansfield: J. Luminesc. **12**, **13**, 271 (1976)
3.36 A. Hutson: Phys. Rev. B **12**, 1404 (1975)
3.37 J. J. Hopfield, H. Kukimoto, P. J. Dean: Phys. Rev. Lett. **27**, 139 (1971)
3.38 R. A. Street, P. J. Wiesner: Phys. Rev. Lett. **34**, 1569 (1975)
3.39 C. H. Henry, M. D. Sturge: Bull. Am. Phys. Soc. **18**, 415 (1973)
3.40 P. J. Dean, R. A. Faulkner: Appl. Phys. Lett. **14**, 210 (1969)
3.41 J. C. Campbell, N. Holonyak, Jr., M. H. Lee, A. B. Kunz: Phys. Rev. B **10**, 1755 (1974)
3.42 M. Altarelli: Phys. Rev. B **11**, 5031 (1975)
3.43 M. H. Lee, N. Holonyak, Jr., J. C. Campbell, W. O. Groves, M. G. Craford, D. L. Keune: Appl. Phys. Lett. **24**, 310 (1974)
3.44 J. J. Coleman, N. Holonyak, Jr., M. J. Ludowise, A. B. Kunz, M. Altarelli, W. O. Groves, D. L. Keune: Phys. Rev. Lett. **33**, 1566 (1974)
3.45 J. D. Cuthbert, D. G. Thomas: Phys. Rev. **154**, 763 (1967)
3.46 N. Holonyak, Jr., J. C. Campbell, M. H. Lee, J. T. Verdeyen, W. L. Johnson, M. G. Craford, D. Finn: J. Appl. Phys. **44**, 5517 (1973)
3.46a S. Gonda, Y. Makita, S. Maekawa: IEEE Trans. ED-**22**, 712 (1975)
3.46b T. Shimada, Y. Shiraki, K. F. Komatsubara: J. Phys. Chem. Sol., **37**, 315 (1976)
3.47 For example, J. M. Ralston: J. Appl. Phys. **44**, 2635 (1973)
3.48 C. H. Henry, R. Z. Bachrach, N. E. Schumaker: Phys. Rev. B **8**, 4761 (1973)
3.49 P. D. Dapkus, W. H. Hackett, Jr., O. G. Lorimer, R. Z. Bachrach: J. Appl. Phys. **45**, 4920 (1974)
3.50 D. R. Wight, J. C. H. Birbeck, J. W. A. Trussler, M. L. Young: J. Phys. D; Appl. Phys. **6**, 1622 (1973)
3.51 For example, T. Beppu, M. Iwamoto, T. Sekiwa, A. Kasami: Japan. J. Appl. Phys. **13**, 1179 (1974)
3.52 M. L. Young, D. R. Wight: J. Phys. D ; Appl. Phys. **7**, 1824 (1974)
3.53 D. F. Nelson, J. D. Cuthbert, P. J. Dean, D. G. Thomas: Phys. Rev. Lett. **17**, 1262 (1966)
3.54 P. J. Dean, R. A. Faulkner, S. Kimura, M. Ilegems: Phys. Rev. B **4**, 1926 (1971)
3.55 R. Z. Bachrach, P. D. Dapkus, O. G. Lorimer: J. Appl. Phys. **45**, 4971 (1974)

3.56   J. M. Dishman, D. F. Daly, W. P. Knox: J. Appl. Phys. **43**, 4693 (1972)
3.57   E. Fabre, R. N. Bhargava, W. K. Zwicker: J. Elect. Mat. **3**, 409 (1974)
3.58   B. L. Smith, T. J. Hayes, A. R. Peaker, D. R. Wight: Appl. Phys. Lett. **26**, 122 (1975)
3.59   C. T. Sah, L. Forbes, L. I. Rosier, A. F. Tasch, Jr., Sol. State Electron. **13**, 759 (1970)
3.60   H. Kukimoto, C. H. Henry, F. R. Merritt: Phys. Rev. B **7**, 2486 (1973)
3.61   P. J. Dean, C. H. Henry: Phys. Rev. **176**, 928 (1968)
3.62   C. H. Henry, D. V. Lang: Phys. Rev. B **15**, 989 (1977)
3.63   H. G. Grimmeiss, L. A. Ledebo, C. Ovren, T. N. Morgan: Proc. Intern. Conf. Physics of Semiconductors, Stuttgart, 1974 (Teubner, Stuttgart 1974) p. 386
3.64   C. H. Henry, H. Kukimoto, G. L. Miller, F. R. Merritt: Phys. Rev. B **7**, 19 (1973)
3.65   D. V. Lang: J. Appl. Phys. **45**, 3014 (1974)
3.66   A. M. White, P. J. Dean, P. Porteous: J. Appl. Phys. **47**, 3230 (1976)
3.67   G. Lucovsky: Sol. State Comm. **3**, 299 (1965)
3.68   The problem lies with the overall spectral shape of the photoionization cross section well above the threshold. For an allowed electric dipole transition, the shape of the cross section very near threshold is invariably well represented by the Lucovsky form $\sigma \propto \Delta E^{3/2}$ independent of the depth of the trap or vagaries in the semiconductor band structure. Thus, this form is generally appropriate for the derivation of threshold energies as is usually done, see for example
       H. G. Grimmeiss, L. A. Ledebo: J. Phys. C-Sol. State Phys. **8**, 2615 (1975)
3.69   D. V. Lang: J. Appl. Phys. **45**, 3023 (1974)
3.70   N. F. Mott, E. A. Davis, R. A. Street: Phil. Mag. **32**, 961 (1975)
3.71   H. W. Moos: J. Lum. **1, 2**, 106 (1970)
3.72   J. S. Jayson, R. N. Bhargava, R. W. Dixon: J. Appl. Phys. **41**, 4972 (1970)
3.73   One of the best authenticated examples may be the deep Ge_{Ga} donor exciton in GaP, P. J. Dean, W. Schairer, M. Lorenz, T. N. Morgan: J. Luminesc. **9**, 343 (1974)
3.74   M. Lax: Phys. Rev. **119**, 1502 (1960)
3.74a M. Jaros: Private communication (1976)
3.75   See Chaps. 2, 3, 5 and 6 in *Physics of Colour Centres*, ed. by W. Beall Fowler (Academic Press, New York 1968)
3.76   D. V. Lang, L. C. Kimerling: Phys. Rev. Lett. **33**, 489 (1974)
3.77   R. D. Gold, L. R. Weisberg: Sol. State Electron. **7**, 811 (1964)
3.78   M. Ettenberg, C. J. Nuese: J. Appl. Phys. **46**, 2137 (1975)
3.79   W. H. Fonger, C. W. Struck: J. Luminesc. **8**, 452 (1974)
3.80   Very little attention has yet been given to the electronic properties of deep levels in $GaAs_{1-x}P_x$, despite the commercial importance of this system (Table 3.1). Typical commerical material is known to contain concentrations of deep levels large compared to the binary compounds, R. N. Bhargava: Private communication (1974). However, the only published work on photocapacitance in $GaAs_{1-x}P_x$ reveals a rather confusing situation, L. Forbes, R. M. Fogle: Appl. Phys. Lett. **25**, 152 (1974)
3.81   R. N. Bhargava, R. M. Harnack, S. P. Herko, P. C. Mürau, R. J. Seymour: to be published. Copper also introduces *two* acceptor levels in GaAs, see [3.17] and D. V. Lang, R. A. Logan: J. Elect. Mat. **4**, 1053 (1975)
3.82   B. Hamilton, A. R. Peaker, S. Bramwell, W. Harding, D. R. Wight: Appl. Phys. Lett. **26**, 702 (1975)
3.82a B. Hamilton: Private communication (1976)
3.83   G. D. Watkins: Intern. Conf. Defects in Semiconductors, Reading 1972 (Institute of Physics, London) p. 228
3.84   C. H. Henry, K. Nassau, J. W. Shiever: Phys. Rev. B **4**, 2453 (1971)
3.85   I. D. Blenkinsop, W. Harding, D. R. Wight, B. Hamilton, A. R. Peaker: Private communication 1975
3.86   T. Kajimura, K. Aiki, J. Umeda: J. Electrochem. Soc. **122**, 15569 (1975)
3.87   M. Ettenburg: J. Appl. Phys. **45**, 901 (1974)
3.88   T. Suzuki, Y. Matsumoto: Appl. Phys. Lett. **26**, 431 (1975)
3.89   D. R. Wight: Private communication (1975)

130   P. J. Dean

3.90   J. Titchmarsh, G. R. Booker: Private communication (1975)
3.91   In contrast to the lack of data on microscopic deep level centers in $GaAs_{1-x}P_x$ mentioned in Section 3.8, it has long been known that dislocation cores produce a strong local reduction in luminescence efficiency in this commercially important LED host. For example
       G. B. Stringfellow, P. E. Greene: J. Appl. Phys. **40**, 502 (1969)
3.92   R. H. Saul: J. Electrochem. Soc. **118**, 793 (1971)
3.93   G. B. Stringfellow, P. F. Lindquist, T. R. Cass, R. A. Burmeister: J. Elect. Mat. **3**, 497 (1974)
3.94   C. Benoit à la Guillaume: J. Phys. Chem. Sol. **8**, 150 (1959)
3.95   A. S. Jordan, A. R. Von Neida, R. Caruso, M. DiDomenico, Jr.: Appl. Phys. Lett. **19**, 394 (1971)
3.96   T. Iizuka: J. Electrochem. Soc. **118**, 1190 (1971)
3.97   R. A. Logan, P. J. Dean, G. Kaminsky: J. Appl. Phys. **42**, 1238 (1971)
3.98   A. S. Barker, Jr., R. Berman, H. W. Verleur: J. Phys. Chem. Sol. **34**, 123 (1973)
3.99   For example the spectrum of Li-diffused GaP:Cd,O shown in Fig. 9 of P. J. Dean: J. Luminesc. **7**, 51 (1973)
3.100 H. Kukimoto, M. Mizuta: Supp. Japan. J. Appl. Phys. **43**, 95 (1974)
3.101 G. A. Rozgonyi, T. Iizuka, S. E. Haszko: J. Electrochem. Soc. **118**, 74C (1971)
3.102 A. S. Jordan, R. Caruso, A. R. Von Neida, M. E. Weiner: J. Appl. Phys. **45**, 3472 (1974)
3.103 J. A. Van Vechten: J. Electrochem. Soc. **122**, 419, 423 (1975)
3.104 Recent evidence for GaAs is cited by S. Y. Chiang, G. L. Pearson: J. Appl. Phys. **46**, 2986 (1975)
3.105 S. Y. Chiang, G. L. Pearson: J. Appl. Phys. **46**, 2986 (1975)
3.106 F. Thompson, S. R. Morrison, R. C. Newman: Intern. Conf. Defects in Semiconductors, Reading, 1972 (Institute of Physics, London) p. 371
3.107 R. C. Newman: Private communication (1975)
3.108 G. D. Watkins: Phys. Rev. Lett. **33**, 223 (1974)
3.109 J. B. Van de Sande, E. T. Peters: J. Appl. Phys. **45**, 1298 (1974)
3.110 D. D. Sell, H. C. Casey, Jr.: J. Appl. Phys. **45**, 800 (1974)
3.111 K. Itoh, M. Inoue, I. Teramoto: IEEE J. QE-**11**, 421 (1975)
3.112 J. A. Rossi, S. R. Chinn, J. J. Hsieh, M. C. Finn: J. Appl. Phys. **45**, 5383 (1974)
3.113 H. Nakashima, N. Chinone, R. Ito: J. Appl. Phys. **46**, 3092 (1975)
3.114 B. C. DeLoach, Jr., B. W. Hakki, R. L. Hartmann, L. A. D'Asaro: Proc. IEEE **61**, 1042 (1973)
3.115 P. Petroff, R. L. Hartman: J. Appl. Phys. **45**, 3899 (1974)
3.116 P. W. Hutchinson, P. S. Dobson, S. O'Hara, D. H. Newman: Appl. Phys. Lett. **26**, 250 (1975); P. W. Hutchinson, P. S. Dobson: Phil. Mag. **32**, 745 (1975)
3.117 R. Ito, H. Nakashima, O. Nakada: Japan. J. Appl. Phys. **13**, 1321 (1974)
3.118 B. W. Hakki, F. R. Nash: J. Appl. Phys. **45**, 3907 (1974)
3.119 H. Kressel, I. Ladany: RCA Review **36**, 230 (1975)
3.120 H. Kressel, H. F. Lockwood, F. Z. Hawrylo: J. Appl. Phys. **43**, 561 (1972)
3.121 Improved life resulting from the addition of small quantities of Al to the active layer may have more to do with the consequent reduction in interfacial contamination by O than the small reduction in interfacial strain due to the lattice mismatch across the metallurgical junction – Y. Nannichi: Private communication (1975)
3.122 W. D. Johnson, Jr., B. I. Miller: Appl. Phys. Lett. **23**, 192 (1973)
3.123 G. A. Rozgonyi, M. B. Panish: Appl. Phys. Lett. **23**, 533 (1973)
3.124 R. L. Hartman, R. W. Dixon: Appl. Phys. Lett. **26**, 239 (1974)
3.125 Y. Nannichi: Private communication (1975)
3.126 J. Matsui, K. Ishida, Y. Nannichi: Japan. J. Appl. Phys. **14**, 1555 (1975)
3.127 T. L. Paoli, B. W. Hakki: J. Appl. Phys. **44**, 4108, 4113 (1973)
3.128 H. Yonezu, M. Ueno, T. Kamejima, I. Sakuma: Japan. J. Appl. Phys. **13**, 835 (1974)
3.129 Y. Nannichi, J. Matsui, K. Ishida: Japan. J. Appl. Phys. **14**, 1561 (1975)
3.130 Created by forward bias and thermal stress according to P. Petroff, O. E. Lorimer, J. M. Ralston: J. Appl. Phys. **47**, 1583 (1976); whereas mechanical stress is employed by M. Iwamoto, A. Kasami: Appl. Phys. Lett. **28**, 591 (1976)

3.131 F. K. Reinhart, R. A. Logan: J. Appl. Phys. **44**, 3171 (1973)
3.132 H. Kressel, I. Ladany, M. Ettenberg, H. F. Lockwood: IEDM Washington 1975, Extended Abstracts,
       A. S. Jordon, A. R. Von Neida, R. Caruso, M. DiDomenico, Jr.: Appl. Phys. Lett. **19**, 394 (1971)
3.133 N. F. Mott, M. Pepper, S. Pollitt, R. H. Wallis, C. J. Adkins: Pro. Roy. Soc. A **345**, 169 (1975)
3.134 N. F. Mott: Electron. Power **19**, 321 (1973)
3.135 R. A. Stradling, R. A. Tidey: Crit. Rev. Sol. State Sci. **5**, 359 (1975)
3.136 R. Dingle, W. Wiegmann, C. H. Henry: Phys. Rev. Lett. **33**, 827 (1974)
3.137 I. Hayashi, M. B. Panish, P. W. Foy, S. Sumski: J. Appl. Phys. **17**, 109 (1970)
3.138 R. Dingle, A. C. Gossard, W. Wiegmann: Phys. Rev. Lett. **34**, 1327 (1975)
3.139 L. Esaki, L. L. Chang: Phys. Rev. Lett. **33**, 495 (1974)
3.140 J. P. vander Ziel, R. Dingle, R. C. Miller, W. Wiegmann, W. A. Nordland, Jr.: Appl. Phys. Lett. **26**, 463 (1975)
3.141 A. Y. Cho, H. C. Casey, Jr.: Appl. Phys. Lett. **25**, 288 (1974)
3.142 H. P. Maruska, J. J. Tietjen: Appl. Phys. Lett. **15**, 327 (1969)
3.143 R. B. Zetterstrom: J. Mat. Sci. **5**, 1102 (1970)
3.144 R. Dingle, D. D. Sell, S. E. Stokowski, M. Ilegems: Phys. Rev. B **4**, 1211 (1971); A. Shintani, S. Minagawa: J. Cryst. Growth **22**, 1 (1974)
3.145 J. I. Pankove, E. A. Miller, D. Richman, J. E. Berkeyheiser: J. Luminesc. **4**, 63 (1971)
3.146 H. P. Maruska, D. A. Stevenson, J. I. Pankove: Appl. Phys. Lett. **22**, 303 (1973)
3.147 J. I. Pankove, M. T. Duffy, E. A. Miller, J. E. Berkeyheiser: J. Luminesc. **8**, 89 (1973)
3.148 D. Wickenden: Private communication (1972)
3.149 M. Ilegems, H. C. Montgomery: J. Phys. Chem. Solids **34**, 885 (1973)
3.150 J. I. Pankove, H. P. Maruska, J. E. Berkeyheiser: Appl. Phys. Lett. **17**, 197 (1970)
3.151 C. D. Thurmond, R. A. Logan: J. Electrochem. Soc. **119**, 622 (1972)
3.152 G. Jacob, R. Madar, J. Hallais: Mat. Res. Bull. **11**, 445 (1976)
3.153 J. P. Dismukes, Y. M. Yim, J. J. Tietjen, R. E. Novak: RCA Review **31**, 680 (1970)
3.154 J. C. Phillips, J. A. Van Vechten: Phys. Rev. Lett. **22**, 705 (1969);
       J. C. Phillips: Rev. Mod. Phys. **42**, 317 (1970);
       J. I. Pankove, H. P. Maruska, J. E. Berkeyheiser: Appl. Phys. Lett. **17**, 197 (1970)
3.155 The optical properties of GaN have been reviewed very recently; J. I. Pankove, S. Bloom, G. Harbeke: RCA Review **36**, 163 (1975)
3.156 R. Dingle, M. Ilegems: Sol. State Commun. **9**, 175 (1971)
3.157 M. Ilegems, R. Dingle: J. Appl. Phys. **44**, 4234 (1973)
3.158 O. Lagerstedt, B. Monemar: J. Appl. Phys. **45**, 2266 (1974)
3.159 M. Illegems, R. Dingle, R. A. Logan: J. Appl. Phys. **43**, 3797 (1972)
3.160 J. I. Pankove, J. E. Berkeyheiser, E. A. Miller: J. Appl. Phys. **45**, 1280 (1974)
3.161 T. Matsumoto, M. Sano, M. Aoki: Japan. J. Appl. Phys. **13**, 373 (1974)
3.162 J. I. Pankove, J. E. Berkeyheiser: J. Appl. Phys. **45**, 3892 (1974)
3.163 J. I. Pankove, E. A. Miller, J. E. Berkeyheiser: J. Luminesc. **5**, 84 (1972)
3.164 G. Jacob: Private communication (1976)
3.165 J. I. Pankove: J. Luminesc. **7**, 114 (1973)
3.166 J. I. Pankove, E. R. Levin: J. Appl. Phys. **46**, 1647 (1975)
3.167 J. I. Pankove, M. A. Lampert: Phys. Rev. Lett. **33**, 361 (1974);
       H. P. Maruska, D. A. Stevenson: Sol. State Elect. **17**, 1171 (1974)
3.168 A. G. Chynoweth, G. L. Pearson: J. Appl. Phys. **29**, 1103 (1958)
3.169 J. I. Pankove: Phys. Rev. Lett. **34**, 809 (1975)
3.170 J. W. Allen: J. Luminesc. **7**, 228 (1973)
3.171 J. I. Pankove, H. Schade: Appl. Phys. Lett. **25**, 53 (1974)
3.172 R. U. Martinelli, J. I. Pankove: Appl. Phys. Lett. **25**, 549 (1974)
3.173 J. I. Pankove: IEEE Trans. ED-**22**, 721 (1975)
3.174 K. Era, S. Shionoya, Y. Washizawa: J. Phys. Chem. Solids. **29**, 1827 (1968);
       K. Era, S. Shionoya, Y. Washizawa, H. Ohmatsu: ibid **29**, 1843 (1968)
3.175 H. Katayama, S. Oda, H. Kukimoto: Appl. Phys. Lett. **27**, 697 (1975)

3.176 T. Inoguchi, M. Takeda, Y. Kakihara, Y. Nakata, M. Yoshida: Digest 1974 SID Intern. Symp. (Winter, New York) p. 84

3.177 A. Vecht, N. J. Werring, R. Ellis, P. J. F. Smith: Proc. IEEE **61**, 902 (1973)

3.178 C. H. Henry: Private communication (1975)

3.179 G. E. Stillman, C. M. Wolfe, J. O. Dimmock: Sol. State Commun. **7**, 921 (1969)

3.180 N. Lee, D. M. Larsen, B. Lax: J. Phys. Chem. Sol. **34**, 1059 (1973)

3.181 R. F. Kirkman, R. A. Stradling: Private communication (1975)

3.182 H. Kressel, H. F. Lockwood, J. K. Butler: J. Appl. Phys. **44**, 4095 (1973)

3.183 W. Heinke, H. J. Queisser: Phys. Rev. Lett. **33**, 1082 (1974)

3.184 R. J. Nelson, N. Holonyak, Jr., W. O. Groves: Phys. Rev. B **13**, 5415 (1976)

3.185 R. J. Nelson, N. Holonyak, Jr., J. J. Coleman, D. Lazarus, W. O. Groves, D. L. Keune, M. G. Craford, D. J. Wolford, B. G. Streetman: Phys. Rev. B **14**, 685 (1976)

3.186 W. Y. Hsu: Bull. Am. Phys. Soc. **22**, 41 (1977)

3.187 R. A. Street, W. Senske: Phys. Rev. Lett. **37**, 1292 (1976)

3.188 D. D. Manchon, Jr., P. J. Dean: Proc. Inst. Conf. Phys. Semicond., Cambridge, Mass., 1970, edited by S. P. Keller, J. C. Hensel, F. Stern, CONF-700801 (Tech. Inf. Serv. Springfield, Va., 1970) p. 760

3.189 A. Baldereschi, N. O. Lipari: Phys. Rev. B **8**, 2697 (1973); ibid. B **9**, 1525 (1974)

3.190 J. I. Pankove, J. A. Hutchby: J. Appl. Phys. **47**, 5387 (1976)

3.191 G. Jacob, D. Bois: Appl. Phys. Lett. **30**, 412 (1977)

4. Recent Advances in Injection Luminescence in II-VI Compounds

Y. S. Park and B. K. Shin

With 16 Figures

Light-emitting diodes (LEDs), diode arrays, and phosphor display panels are finding increased use in a variety of commercial applications. Present and anticipated applications of these devices include solid state indicators (e.g., digital clocks, meter readouts) and display systems (e.g., instrument panels, TV displays), the application being determined by the light-output capability and size availability (cost) of the particular device.

In order to achieve flexibility and versatility in display systems, LEDs over the entire visible spectrum are needed. Current efforts in the development of such devices, therefore, are directed toward improving efficiency and increasing the number of available colors. Improvement of LEDs also depends critically upon the preparation, evaluation, and characterization of new and improved diode substrate materials. In order to qualify as an efficient visible LED material, a semiconductor must have a wide bandgap ($\gtrsim 1.8$ eV), produce stable pn junctions, and provide favorable radiative recombination transitions.

For many years wide bandgap II-VI compounds of the Zn and Cd chalcogenides (such as CdS, ZnTe, ZnSe, and ZnS) have been of great interest to the scientific community as potential visible LED materials since they possess excellent luminescent properties and have bandgaps which range throughout the visible spectrum into the near ultraviolet (UV), producing high-efficiency luminescence under UV and electron-beam excitation. These materials can also be produced in the form of mixed-crystal (alloy) combinations such as CdTe − ZnTe, ZnTe − ZnSe, ZnSe − ZnS, and ZnS − CdS in varying proportions to achieve a continuous luminescence spectrum in the visible range. Furthermore, these are direct bandgap materials and, therefore, an efficient radiative recombination can be achieved in a straightforward manner − i.e., band-to-band transitions − since no crystal-momentum-preservation problems exist.

However, in the past, much of this potential has not been realized due to the difficulty in achieving low-resistivity amphoteric doping (pn junctions) required for efficient carrier injection and, in turn, efficient radiative recombination. This difficulty in producing amphoteric doping in II-VI compounds is due to compensation of native defects during crystal growth, low uncompensated solubility of intentionally incorporated impurities, and high activation energies of known donor or acceptor states.

As a result, only CdTe ($E_g = 1.5$ eV) can be made in both low-resistivity n- and p-type form; CdS, ZnSe, and ZnS can be made only in low-resistivity n-type form; and ZnTe can be made only in low-resistivity p-type form by conventional doping methods. Although gallant efforts have been made in the past to improve the doping process, very little success has been achieved. It must be pointed out that research efforts on II-VI compounds have been quite restricted in comparison to those on III-V compounds. As a matter of fact, research efforts on all of the II-VI materials combined have been less than those on GaAs alone. As a result, progress on II-VI compound technology at present is at the stage where GaAs technology was a decade ago. Recent advances in the field of injection electroluminescence should stimulate the interest of both the scientific and industrial community in II-VI compounds.

In the past decade or so, tremendous progress has been made in the fabrication of III-V compound LEDs; the performance of recently made red-light-emitting alloy diodes made from Ga(As, P), for example, is unsurpassed and no contest is expected from II-VI compound materials in this spectral region. In the yellow-green and green spectral regions, great advances have been made in GaP:N and (Ga, Al)P materials; the II-VI counterparts of these materials (ZnTe, for example) may not be able to compete with these III-V materials. However, in the deep green ($\lambda \lesssim 5400$ Å) to blue spectral region, efficient injection-luminescent devices have not been obtained. Luminescence in this spectral region would require materials having a wider bandgap such as SiC and GaN, as well as CdS, ZnSe, and ZnS from II-VI. CdS was one of the first II-VI systems investigated for injection luminescence. However, this material failed to produce even fairly efficient room-temperature emission in the visible spectral region.

Presently, much progress is being made on blue LED efforts in GaN; the most likely candidate for blue emission in II-VI compounds being the ZnSe or ZnS system. Persistent efforts to produce LEDs from II-VI materials stem from the realization that injection luminescence with a brightness of $> 10^4$ fL should be feasible in the blue spectral region (~ 4400 Å) with ~ 2.8 V bias voltage at 1 A/cm^2 if a proper method (*pn* junction) of efficient carrier injection can be provided. Recent developments in both sample preparation and device fabrication through the use of ion implantation doping and improved diffusion doping are encouraging. Previous efforts in injection electroluminescence describing the mechanisms of carrier injection are well documented in a number of review articles [4.1–5]. Recently, an excellent review on the subject was presented by *Aven* and *Devine* [4.6], their discussions centering mainly around ZnS and ZnSe$_x$Te$_{1-x}$ systems. Discussions concerning more recent developments in injection luminescence from metal-semiconductor, metal-insulator-semiconductor, and heterojunction structures as well as alloy and *pn* junctions formed by ion implantation have not been considered.

In view of this fact, this chapter will contain an overview of more recent developments in injection electroluminescence in II-VI compounds and a brief discussion of LED structures employed. This discussion will follow a brief review of material properties and luminescence transitions, emphasizing electrically and optically active impurity centers for various material systems. Finally, a brief assessment of the future of II-VI materials will be given.

4.1 Properties of II-VI Compounds

Each semiconducting material has its own unique physical, chemical, electrical, and optical characteristics. Properties of II-VI compounds have been reviewed previously [4.1, 2]. In this chapter, discussion will be limited to those properties which are directly related to injection-luminescence effects in II-VI compounds. Furthermore, the discussion will be limited to CdS, ZnTe, ZnSe, ZnS, and their alloys since present interest involves the visible luminescence, and other II-VI materials are not suitable for this application.

Physical and optical properties such as crystal structure, lattice constant, energy gap, refractive index, and lasing line which are important in the study of injection-luminescent devices are given in Table 4.1. Electronic properties of II-VI compounds relevant to this discussion are given in Table 4.2.

Table 4.1. Some properties of II-VI compounds

Properties	CdS	ZnTe	ZnSe	ZnS
Crystal structure	Z, W	Z	Z, W	Z, W
Lattice constant (A)	5.464 (Z) a = 4.1368 (W) c = 6.7167 (W)	6.103 (Z)	5.668 (Z) a = 4.01 (W) c = 6.54 (W)	5.409 (Z) a = 3.806 (4 H) c = 12.44 (4 H)
Thermal expansion coefficients (~ 300 K) (10^{-6}/K)	4	8.5	7.2	6.14
Ionicity of bonds (%)	22	9	15	24
Energy gap (300 K) (eV)	2.38	2.26	2.67	3.66
Refractive index	2.506 (6000 Å)	3.165 (5200 Å)	2.61 (5890 Å)	2.49 (4400 Å)
Lasing line (Å)	4897 (4.2 K) 4900 (110 K) 5270 (300 K)	5280 (4.2 K) 5310 (77 K) 5330 (110 K)	4600 (150 K)	3290 (4.2 K)
Max melting point (°C)	1475	1295	1520	1830

Table 4.2. Some electronic properties of II-VI compounds

Electronic properties	CdS	ZnTe	ZnSe	ZnS
Dielectric constant				
static (ε_0)	10.3	10.4	9.2	8.9
optical (ε_∞)	5.4	7.3	6.1	5.7
Mobility (cm^2/V s)				
electron (μ_n)	250	340	600	160
hole (μ_p)	15	110	40	10
Effective mass				
electron (m_e^*/m_0)	0.208	0.09	0.17	0.34
hole (m_h^*/m_0)	0.80	0.15 (l_p)	0.60	0.58
		0.68 (h_p)		
Work function (ϕ) (eV)	5.01	5.43	4.84	5.40
Electron affinity (χ) (eV)	4.79	3.53	4.09	3.90

4.1.1 II-VI Compounds and Injection Electroluminescence

II-VI compounds of CdS, ZnTe, ZnSe, and ZnS have a wide energy bandgap between 2.26 and 3.66 eV. The most frequently used criterion in classifying materials used in injection electroluminescence is their band structure. II-VI compounds have direct bandgaps, which means that the energy minimum occurs at $k=0$ in the energy-momentum ($E-k$) relationship, an example of which is given in Fig. 4.1 for a ZnSe crystal [4.7]. Therefore, in the Cd and Zn

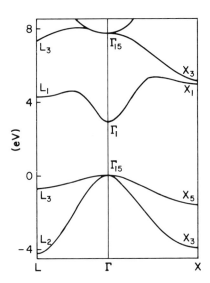

Fig. 4.1. Energy band structure of ZnSe

chalcogenides crystal momentum conservation is not a problem and, as a result, there is an efficient, straightforward, band-to-band radiative recombination. This means that efficient electroluminescence and also lasing action due to carrier injection can be expected in the visible region. For example, near bandgap luminescence of green ($\lambda \simeq 5\,500$ Å) from ZnTe, deep green ($\lambda \simeq 5\,200$ Å) from CdS, blue ($\lambda \simeq 4\,600$ Å) from ZnSe, and UV ($\lambda \simeq 3\,400$ Å) from ZnS can be obtained (see Fig. 4.2). A comparable spectral range in the III-V compounds will be yellow-green to green ($\lambda \simeq 5\,500$–$5\,650$ Å) from GaP:N and (Ga, Al)P, green to deep green ($\lambda \simeq 4\,800$–$5\,400$ Å) from GaAs with up-converting phosphors, and blue to near UV ($\lambda \simeq 3\,700$–$4\,400$ Å) from GaN. In the II-VI compounds ZnTe diodes must compete with well-advanced GaP and Ga(As, P) since they cover the same spectral range. In earlier days, CdS showed great promise for producing green luminescence. However, to date efficient injection luminescence has not been obtained at room temperature. This leaves only two systems – ZnSe and ZnS – in II-VI compounds for potential blue-electroluminescent devices. The development of semiconducting materials for blue-electroluminescent devices is generally complicated by either the crystal growth conditions (e.g., SiC) or the difficulty in achieving p-type doping (e.g., in ZnSe, ZnS, and GaN). In the case of ZnSe which has a "blue" band gap ($E_g = 2.67$ eV), the difficulty lies in obtaining low-resistivity p- and n-type samples which emit light in the band-edge region. Furthermore, for UV bandgap compounds

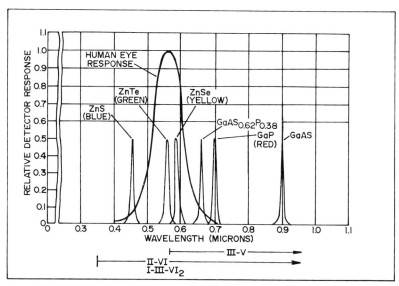

COMPARISON OF II-VI AND III-V LEDs
I-III-VI$_2$

Fig. 4.2. Comparison of II-VI, III-V, and I-III-VI$_2$ optical spectral ranges

(SiC, ZnS, and GaN), it is necessary to determine the doping conditions which involve deep centers in order to obtain efficient room-temperature blue luminescence. In the past several years, encouraging developments have taken place in both ZnSe and ZnS materials – likely candidates for future "blue" LEDs.

4.1.2 Material Considerations

The II-VI family exhibits properties which are intermediate between the covalent group IV elements and the III-V compounds on the one hand and the ionic I-VII compounds on the other. The covalent series – Ge, Si, and diamond —demonstrates that decreasing atomic size and lattice constants lead to increasing bandgap and crystal cohesion energy. On the other hand, in the isoelectronic series – Ge, GaAs, and ZnSe – increasing electronegativity (or ionicity of the bonds) at constant lattice parameters leads to increasing bandgap but decreasing cohesion energy. The wide bandgap II-VI compounds – CdS, ZnSe, and ZnS—stem from a contribution of increasing ionicity in mixed covalent-heterovalent bonds, and the bandgaps tend to increase with decreasing atomic size.

Almost all of the II-VI compounds can be prepared either in a hexagonal (wurtzite) structure or as cubic (zincblende) crystals. The interatomic distances are practically the same for compounds having both forms [4.8], which implies that their local environments and chemical bonding are nearly identical. Furthermore, their band structures, which can be partially inferred from those of cubic crystals by means of a perturbation approach [4.9], are not expected to be grossly different.

Single crystals of II-VI compounds—CdS, ZnTe, ZnSe, and ZnS—can be grown by vapor or chemical transport techniques and from the melt. The methods used to grow particular systems vary the crystal structure and defect formation which, in turn, affects electrical and optical properties of the material. The first sizable synthetic crystals of II-VI compounds were grown from the vapor phase [4.10]. This method has been used with considerable success in the growth of large high-purity single-crystal CdS platelets [4.11]. For the preparation of large-volume single crystals, growth from the melt has been used primarily; materials prepared by this technique are normally found in the cubic (zincblende) form, but samples grown from the vapor phase have shown that wurtzite (hexagonal)-type crystals, although they contain intergrowths of the zincblende type. For example, melt-grown hexagonal ZnS crystals contain a considerable degree of crystallographic disorder, whereas cubic crystals are crystallographically much better defined.

Alloys of II-VI compounds can be prepared from combinations such as ZnS – ZnSe, ZnSe – ZnTe, ZnTe – CdTe, and ZnS – CdS in all proportions in such a way that continuous bandgap values can be realized over the entire visible spectrum. Of the II-VI compounds of interest, the ZnS – ZnTe and the CdS – CdTe mixed crystal systems are the only ones not completely miscible as solid solutions. Variation of bandgap with composition for $Zn_xCd_{1-x}Te$, $ZnSe_xTe_{1-x}$, and $Cd_{1-x}Zn_xS$ three-component systems is shown in Fig. 4.3. With the exception of $ZnSe_xTe_{1-x}$, the bandgap varies monotonically with composition for the $Cd_{1-x}Zn_xS$ system [4.12] and somewhat less for the $Zn_xCd_{1-x}Te$ system [4.13]. The bandgap increases linearly with S concentration for the ZnS_xSe_{1-x} system [4.14] (not shown in the figure) from 2.67 eV for ZnSe up to 3.66 eV for ZnS. In the $ZnSe_xTe_{1-x}$ system [4.15] the bandgap does not increase monotonically with Se concentration but has a broad minimum around $x \simeq 0.4$ which is somewhat lower than 2.26 eV. This anomalous variation of bandgap with composition has attracted much attention as a possibility for obtaining efficient injection luminescence.

Because of the difficulty in obtaining pn homojunctions in II-VI compounds, these alloy systems have attracted much attention for obtaining pseudo-binaries for the preparation of efficient injection luminescent devices. Among the alloy systems only $Zn_xCd_{1-x}Te$, $ZnSe_xTe_{1-x}$, and ZnS_xSe_{1-x} can be amphoterically prepared with a bandgap in the visible range. CdTe can be prepared degenerately in the n-type and low-resistivity p-type forms, while ZnTe can be made only low-resistivity p-type. Crystals of the composition $Zn_xCd_{1-x}Te$ can be doped both n- and p-type for an experimental value [4.16] of $x \simeq 0.7$. For $ZnSe_xTe_{1-x}$, ZnSe is normally prepared in the low-resistivity n-type form and ZnTe in the low-resistivity p-type form. In this system, in order to form a suitable pn junction [4.17], the crystals can be prepared both

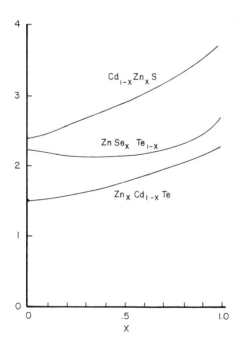

Fig. 4.3. Variation of bandgap with composition in mixed crystals of II-VI compounds

n- and p-type for x in the range $0.35 \lesssim x \lesssim 0.5$. Recently, a pn junction injection luminescent device was prepared from the ZnS_xSe_{1-x} system [4.18]. Since both ZnSe and ZnS can be prepared only in the low-resistivity n-type form, pn junctions from this system ordinarily would not be expected. However, from this unusual system, a p-type layer was prepared using the diffusion process from a Ga or In source with a $ZnS_{0.45}Se_{0.55}$ crystal [4.19][1]. Diodes prepared from various composition and diffusion ranges yielded injection luminescence in the yellow, yellow-green, green, and blue-green spectral regions. More details on this system will be given in the following sections.

4.1.3 Impurity Doping

Impurities are incorporated into semiconductors in order to achieve electrically and optically active centers. Some of the well-known donor and acceptor impurities in II-VI compounds are listed in Table 4.3. In II-VI compounds, either substitutional metal or chalcogenide sites must be considered. The rule of thumb [4.20] for associating the ionization level with a particular sublattice site is related to the radius of the site; in particular, the larger potential metal and chalcogenide sites would lead to a deeper center (radius of cation and anion for CdS, ZnTe, ZnSe, and ZnS is given in Table 4.4). As an example, a Group V element occupying a substitutional anion site would appear to be a reasonable choice for acceptor doping in CdS, ZnSe, and ZnS, which satisfies the requirement for minimizing self-compensation (a deep impurity level).

The behavior of group III elements such as Al, Ga, and In is well established—they substitute for Zn or Cd and act as donors. In CdS, low-resistivity n-type samples can be prepared from these elements with little difficulty. In the Zn Compounds these donors are largely compensated by Zn vacancies, and as-grown bulk materials exhibit high resistivity. The role of halogens (F, Cl, Br, I) is quite similar to that of group III donors in that the halogen donors form a complex in which they associate with a Zn or Cd vacancy. In the case of F, electrical measurements [4.21] have shown that besides acting as a donor, it also introduces acceptor states near the conduction band in CdS, presumably by trapping anion vacancies. II-VI materials are often prepared by firing in the alkali halides (e.g., Li, Na) and are frequently seen in the 1–100 ppm range

---

[1] From a practical point of view, the efficient generation of light by minority carrier injection in II-VI compounds is a great accomplishment regardless of whether the holes come from a p-type region or from a Schottky barrier. However, the evidence for a p-type layer must be interpreted with caution. All the claims for a p-type layer in II-VI compounds were made always for very thin layers. It is difficult to distinguish between an inversion layer and an equally thin p-type layer. Channeling effect may join the Schottky barrier contacts used in Hall effect and conductivity measurements and give a p-type reading. In fact, a field induced p-type channel has been proposed as a source of holes for a CdS LED by Goodman in RCA Review 34, 429 (1973). Hence, although the distinction between p-type and inversion layer may be academic in the case of thin regions, this possibility of confusion must be pointed out.

Table 4.3. Donor and acceptor impurities in II-VI compounds

Group	Impurity	Sub. for	Donor	Acceptor
IIIA	B, Al, Ga, In, Tl	Zn, Cd	D	
VIIA	F, Cl, Br, I	S, Se, Te	D	
IA	Li, Na	Zn, Cd		A
IB	Cu, Ag, Au	Zn, Cd		A
VA	N, P, As, Sb, Bi	S, Se, Te		A

Table 4.4. Nearest bond distance and tetrahedral radius for some of II-VI compounds

	Tetrahedral radius (Å)		Nearest bond distance (Å)	r_c/r_a
	Cation (r_c)	Anion (r_a)		
CdS	1.48	1.04	2.52	1.42
ZnTe	1.31	1.32	2.63	0.99
ZnSe	1.31	1.14	2.45	1.15
ZnS	1.31	1.04	2.35	1.26

in mass-spectroscopic analysis of single crystals, but the role of these elements is not well understood. For example, the behavior of Li in ZnSe is suspected to be much the same as in GaAs [4.22]; Li can occupy both substitutional and interstitial sites, and GaAs is isoelectronic with ZnSe having the same crystal structure and nearly the same lattice constants. When substituted for a Zn or Cd sublattice, these alkali metals are expected to act as p-type impurities. Group I (Cu, Ag, and Au) elements are known to diffuse extremely rapidly as interstitials, but they can also be incorporated into the lattice as substitutes for Zn or Cd. Although in most cases these elements form complexes with native defects or other impurities, they may also be incorporated as isolated substitutional acceptors in Zn chalcogenides. Group V (N, P, As, Sb, and Bi) elements are known to occupy an anion-sublattice position and act as acceptors. However, these elements have very low solubility and form deep acceptor centers. As an example energy levels and electrical properties of known impurities in ZnSe are listed in Table 4.5.

Table 4.5. Energy levels and transport properties of known impurities in ZnSe

Dopant	Conductivity type	E_A or E_D	ρ (Ω-cm)	μ_h (cm^2/V s)	N (cm^{-3})	Ref.
Li	p	0.66	$2 \cdot 10^8$	23	$2.7 \cdot 10^9$	4.23
Li[a]	p	0.114	—	—	—	4.24
Na[b]	p	0.085–0.100		7.2		4.25
Na[a]	p	0.09				4.24
Cu	p	0.072				4.26
Cu	p	—	—	28	—	4.27
N[b a]	p	0.136				4.25
P	p	0.68	$1.2 \cdot 10^8$	24	$4 \cdot 10^9$	4.28
P[b]	p	shallow	10	6	10^{17}	4.29
I$_{Se}$	p	0.68				4.30
V$_{Zn}$	p	~ 0.1	$8 \cdot 10^7$	25	$5 \cdot 10^9$	4.31
Ga, In	p		20	2.7	$1.9 \cdot 10^{17}$	4.18
Al[a]	n	0.026				4.32
Al	n		0.74	365	$2.6 \cdot 10^{16}$	4.34
Ga[a]	n	0.028				4.32
Ga	n		0.2	410	$8.5 \cdot 10^{16}$	4.34
In[a]	n	0.029				4.32
F[a]	n	0.029				4.32
Cl[a]	n	0.027				4.32
Cl	n		0.3	550	$6.8 \cdot 10^{16}$	4.35

[a] Optical measurements.
[b] By ion implantation.

4.1.4 Compensation Effect

Electrical compensation effects alluded to in the previous discussions will now be considered in more detail. In II-VI compounds, it is difficult to obtain amphoteric doping by conventional crystal-growth methods. As a result only CdTe can be made in both the low-resistivity n- and p-type form, ZnTe can be made only in the low-resistivity p-type form, and the remaining II-VI compounds can be made only in the low-resistivity n-type form. The failure of amphoteric doping is primarily due to electrical compensation of the intentionally added donor or acceptor impurities by native defect centers of the opposite conductivity type. The mechanism for self-compensation may be thought of in terms of a balance of energies. Energy is required to produce lattice defects such as vacancies which act as compensating centers, whereas energy is gained by the lattice through the interaction of these defects with free carriers available from the impurity centers introduced. Thus, if the energy required to form defects is large compared to the energy gained by compensation, the compensation of impurity centers by the defects becomes unavoidable. The defect formation energy is a function of both lattice temperature and bandgap; in wide-bandgap materials at elevated temperatures, compensating centers are formed with relative ease. The energy released to the system through compensation is related to the ionization energy of the acceptor center being compensated; therefore, the closer the acceptor level is to the valence band, the more easily compensation can occur. For a given material, the probability of self-compensation can, therefore, be minimized by choosing a deep acceptor and introducing it at low temperatures.

While there is considerable experimental evidence to support the premise that compensation by native defect centers is a predominant factor in the failure to achieve amphoteric doping, additional factors are also important. For example, the solubility limit of some otherwise suitable shallow acceptor impurities in sulfides and selenides may be too low for compensation of the donor impurities accidentally introduced during the crystal growth. There is evidence that some other acceptor impurities such as Ag in CdS [4.36], Li in CdS [4.37], and Li in ZnTe [4.38] may undergo self-compensation by distributing themselves between donor and acceptor sites. The dopants may also form electrically and optically inactive complexes with impurities or native defects unintentionally introduced or created in the material during the growth process. Several halogens have been found to form neutral complexes such as Cl with Sr in ZnSe [4.39]. Finally, most II-VI compounds contain a high concentration of interstitial metal and chalcogen atoms. Such interstitials have not been demonstrated to exhibit electrical activity at room temperature, but it is conceivable that the interaction of these interstitials with dopants during diffusion may result in the formation of stable neutral centers. These may occur in the form of atomically dispersed complexes, precipitated second phases which extract the added impurities, or electrically active compensating centers [4.2].

4.1.5 Impurity Doping by Ion Implantation

In the above discussions, it has been pointed out that the major difficulties in achieving amphoteric doping involve self-compensation and low solubility of dopant impurities. An alternative method of doping which offers at least two major advantages over the conventional doping method is ion implantation. Since this is a non-equilibrium process, the amount of dopant incorporated depends only upon energy and beam current of the ions being implanted. Furthermore, since the doping can be performed at relatively low temperatures, the formation of compensating defect centers can be prevented. Earlier results of ion-implantation efforts in II-VI compounds have demonstrated this to be an effective means of type conversion in such systems as CdTe:As [4.40], CdS:Bi [4.41], CdS:P [4.42, 43], CdS:N [4.44], ZnTe:F [4.45], ZnTe:Cl [4.46], and ZnSe:Li [4.47].

Recently, *Park* and *Shin* [4.29] produced a low-resistivity *p*-type layer in P-ion implanted ZnSe and obtained efficient ZnSe injection electroluminescent diodes [4.48]. The substrate materials used in this investigation were melt-grown single-crystal cubic *n*-type ZnSe heavily doped with Al. Due to defects produced during the growth process, the resistivity of the as-grown crystals was high, normally in the range of 10^7–10^9 Ω-cm at room temperature, although

n-type dopants such as Al might be expected to produce low-resistivity n-type ZnSe. The crystals were subjected to a molten Zn treatment for 24 h at 900 °C by means of a technique described previously by *Aven* and *Woodbury* [4.49]. The resulting heat-treated crystals had measured resistivity values of 0.15 Ω-cm, electron mobility of 400 cm^2/V s, and carrier concentration of $2 \cdot 10^{17}$ cm^{-3} at room temperature. The P-ions were implanted at 400 keV to a dose of 10^{14} ions/cm^2 at room temperature. The implanted samples were subjected to annealing at temperatures up to 500 °C for times up to 1 h in Ar-gas ambient.

Sheet-resistivity and Hall-effect measurements were made on well-annealed samples; since formation of "good" junctions is required for Hall-effect measurements, the presence of a well-defined depletion layer and low junction-leakage currents will electrically isolate the implanted layer from the substrate. Au contacts of 0.1 cm diam were evaporated (or sputtered) onto the implanted layer at 0.5-cm intervals of the van der Pauw pattern. In II-VI compounds, it is difficult to obtain reliable, reproducible low-resistivity ohmic contacts. Good ohmic contacts [4.50] in the p-type samples are metal electrodes which have a larger work function than that of p-type semiconductors and, when diffused into the semiconductor, act as acceptor dopants. In the present case, the contact resistance could be ignored since that of the Au contact on n-type material is $\sim 10^9$ Ω in the voltage range up to ~ 4 V and differs by 10^3 from that of the implanted layer (see Fig. 4.4). Another precaution to be taken in Hall measurements on ion-implanted samples is that the measurements must be made within the ohmic region of the sample current since high applied current may deteriorate junction characteristics and may become nonlinear because of leakage currents through the substrate as a result of barrier break-down which yields erroneous Hall-measurement results. Hall measurements made under these conditions indicate that p-type layers having a low resistivity

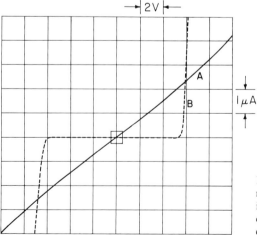

Fig. 4.4. *I-V* characteristics for sheet-resistance measurement: (A) p-type implanted ZnSe, (B) n-type ZnSe. The central square marks the origin of the coordinates

of ~ 10 Ω-cm, a hole mobility of 6 cm^2/V s, and a hole concentration of 10^{17} cm^{-3} were obtained as a result of P-ion implantation in ZnSe. From the concentration of donors, $N_D \simeq 10^{17}$ cm^{-3} and acceptors $N_A \simeq 10^{17}$ cm^{-3}, the diffusion potential of an abrupt junction can be estimated from

$$eV_D \simeq kT \ln\left(\frac{N_A N_D}{n_i^2}\right).$$

This calculation yields a diffusion potential of 2.51 eV at room temperature. In n-type material for $N_D = 10^{17}$ cm^{-3}, the Fermi level is about 0.075 eV below the conduction band edge. The Fermi level of the p-type layer calculated is then 2.585 eV below the conduction band edge or ~ 0.085 eV above the valence band edge at room temperature. This result agrees well with the recent observation of 0.07-eV energy level produced by P-ion implantation in ZnSe [4.51].

Phosphorus is known to form an isolated acceptor center by substituting for selenium in melt-grown ZnSe as identified by electron-paramagnetic-resonance studies [4.52]. Acceptor-type doping has also been reported in a ZnSe system containing Li and Cu [4.23, 26, 27]. Activation energies of these acceptor centers including ZnSe:P [4.28], are reported to be in the range 0.6–0.75 eV from the valence band edge. However, in a recent report on p-type conduction in undoped ZnSe [4.31], a shallow acceptor level (~ 0.1 eV) was measured in addition to a deep level (0.65–0.75 eV). In an ion-implanted II-VI system, a shallow acceptor level of 0.05 eV in P-implanted CdS [4.53] and 0.07 eV in P-implanted ZnSe [4.51] was also measured. These investigators concluded that a complex center is responsible for the formation of the shallow acceptor level. Defect interaction with intentionally doped impurities in II-VI compounds (as compared to other semiconductors such as III-V compounds) is found to be advantageous. Implantation of suitable impurities can be used to produce active electrical and optical centers not possible with material containing only one type of defect center.

More recently, the electrical properties of Na-implanted ZnSe were investigated in detail [4.25]. In CdS, substitutional Na acceptors are known to be compensated by interstitial Na donors [4.54]. Although this characteristic may also be present in ZnSe, the shallow activation energy and known high solubility [4.55] of Na in ZnSe were the primary reasons for recent investigations. Hall-effect measurements on well-annealed samples provided evidence of p-type conductivity in the Na-implanted ZnSe. Hall hole mobilities of up to 7 cm^2/V s, room-temperature resistivity of 15 Ω-cm, and an effective bulk hole concentration of $6 \cdot 10^{17}$ cm^{-3} have been estimated in the p-type layer formed by assuming a uniformly implanted layer of 1 μm. Temperature dependence of the sheet-resistivity measurements indicated an acceptor activation energy of 110 meV which was consistent with the results of photoluminescence measurements.

4.2 Luminescence Transitions

Since an efficient luminescent transition is required for obtaining efficient light emission from injection electroluminescent devices, the mechanism for this transition must be consistent with that for current transport. Therefore, a discussion of the basic mechanisms responsible for common luminescence transitions in II-VI compounds will be given in this section. Luminescence in II-VI compounds generally falls into three categories: exciton transitions, edge and pair emission, and broad-band luminescence. Typical luminescence spectra of II-VI compounds from the ZnSe sample are illustrated in Fig. 4.5.

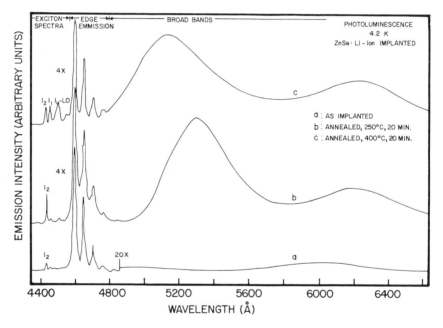

Fig. 4.5. Photoluminescence spectra of a Li-ion-implanted ZnSe at 4.2 K after various annealing stages

The sharp lines having widths of 0.01–1.0 meV are exciton emission; the next series of lines having widths > 10 meV represents edge emission. Historically all luminescence near the band edge energy has been called edge emission, with excitons being a type of edge emission. We will follow the more recent convention of referring to the lower energy transitions having line widths > 10 meV as "edge emission". The broad-band luminescence shown in the figures at low energy is associated with deep centers.

4.2.1 Exciton Transitions

The sharp-line spectrum near the band edge is seen only at low temperatures and is the result of exciton emission. An exciton is made up of a free electron and free hole which are coupled by mutual Coulomb attraction. The electron and hole can orbit around their common center of mass in quasi-hydrogen orbitals. This pair is free to move throughout the crystal lattice until it decays as free exciton emission or until it is trapped at an impurity or defect. The trapped excitons possess modified orbitals which are characteristic of the trapping center. The trapped exciton when it decays, therefore, provides additional information about the center. The ionization energy for the free exciton, relative to the conduction and valence band electron and hole states which are analogous to the free electron states in the hydrogen atom model, is given by

$$E_{ex}^{Free} = \frac{-m_r^* e^4}{2h^2 \varepsilon^2} \frac{1}{N^2} \qquad N = 1, 2, \ldots$$

where

$$\frac{1}{m_r^*} = \frac{1}{m_e^*} + \frac{1}{m_h^*}.$$

m_r^* is the reduced mass, m_e^* and m_h^* are the effective mass of the electron and the hole, respectively, and ε is the dielectric constant of a given material. Since the exciton is free, it is not spatially localized. The energy E_{ex} of an exciton is generally represented relative to the conduction band edge, $E_{ex}(N = \infty)$ being the conduction band energy. When the free exciton recombines, the emitted energy is

$$hv = E_g - E_{ex}^{Free}$$

where E_g is the bandgap energy. This expression neglects the translational kinetic energy of the free exciton, which broadens the observed transition.

In materials which contain defects or impurities, complexes called bound excitons occur. A free exciton may be trapped at the site or a free hole may combine with a neutral (un-ionized) donor (substitutional impurity with an extra electron) to form a positively charged ion. In this case the electron travels about the donor atom in a wide orbit, and the hole moves in the dipole field of the orbiting electron. This behavior is analogous to that of a free electron and neutral acceptor. Bound excitons are characterized by more sharply peaked emission which occurs at a lower energy. A bound exciton may be identified by the Zeeman effect as being bound to a donor or an acceptor. The emitted energy for the bound exciton is

$$hv = E_g - E_{ex}^{Free} - E_{ex}^{B}$$

where E_{ex}^B is the energy binding the exciton to the defect center. Detailed discussions of excitons in II-VI compounds can be found in a number of excellent review articles [4.56].

4.2.2 Edge Emission and Pair Spectra

So-called edge emission can be used to describe radiative recombination processes occurring within a few kT of the band edge. Such processes usually dominate the photoluminescence spectra at low temperatures of ~ 100 K. Historically the term edge emission referred to all luminescent transitions near the band edge, including the excitonic spectra discussed in the previous section. In the II-VI compounds only that radiative emission having energy less than E_g is observed. Optical absorption is extremely high for energies very near the bandgap and, consequently, any conduction-band-to-valence-band direct recombination which occurs radiatively is reabsorbed. The observable edge emission is attributed to free-to-bound transitions from the conduction band to the relatively shallow acceptor levels or from shallow donors to the valence band. In addition, bound-to-bound (donor-acceptor) transitions have also been observed.

In a free-to-bound transition, the radiative recombination occurs when free electrons recombine at neutral acceptors or free holes recombine at neutral donors. The emitted radiation from the free-to-bound transition may be expressed as

$$hv = E_g - E_A \text{ (or } E_D).$$

These transitions are characterized by a line width proportional to the kinetic energy of the free carriers in the conduction (or valence) band which is determined by the thermal energy. This is expressed approximately as

$$H(T) \approx 2.5 \, kT.$$

Luminescence peaks of free-to-bound transitions are relatively independent of excitation intensity. This behavior has been reported by *Thomas* et al. [4.59], in CdS samples. Transitions of this type may quench pair transitions, especially at temperatures of $\gtrsim 50$ K. In CdS the spectral features of the edge emission undergo a considerable change between 4.2 and 77 K. This change is attributed to $D-A$ transitions which give way at higher temperatures to more probable recombination of free electrons with holes at acceptor centers [4.57, 58].

In a donor-acceptor pair recombination, the photon energy emitted from the pair having separation r is given by the following equation, provided the phonon cooperation is absent

$$hv = E_g - \left(E_A + E_D - \frac{e^2}{\varepsilon r} \right),$$

the last term representing the Coulomb interaction between the donor and the acceptor. Since possible values of r are discrete in a crystal lattice, multi-line luminescence spectra characteristics of the impurity-site locations are expected. The transition probability of the pair emission depends upon the interpair separation r. When either donor or acceptor is sufficiently shallow, the transition probability $W(r)$ is given by

$$W(r) = W_0 \exp(-2r/r_B),$$

where W_0 is a constant for a given $D-A$ pair in a semiconductor and r_B is the Bohr radius of the shallower state. As the $D-A$ separation increases, the lines tend to merge into a continuum, and at the same time the lifetime of the state increases steadily. This is an important characteristic of distant $D-A$ recombination. Furthermore, the long lifetime means that this type of luminescence is readily quenched by thermal ionization and is expected only at low temperatures. This type of recombination has been observed in several semiconductors including some II-VI compounds [4.60].

Three experimental approaches may be employed to verify that an emission peak originates from $D-A$ pair recombination. The direct proof is the observation of a series of sharp lines due to the pair recombinations which correspond to the different values of r, as was originally shown in GaP [4.61, 62]. Another method is the observation by time-resolved luminescence measurements [4.63] of an energy shift of the emission peak due to transitions between the relatively distant pairs after pulse excitation. Finally, another evidence of $D-A$ recombination [4.64] is the detection of an energy shift toward the higher-energy side with increasing excitation intensity. This shift has been attributed to saturation of distant $D-A$ pair states having long lifetime and is considered to be a characteristic of $D-A$ recombination.

In binary compound semiconductors, the anions and cations form similar but separate sublattices. Donors and acceptors which occupy the same sublattice are called Type I pairs; if they are not on the same sublattices, they are called Type II pairs. The magnitude of r depends upon the pair type and also upon the doping concentration.

Halsted [4.65][2] has developed an expression to relate the ionization energy of a localized defect or impurity state to the ionization energy of the exciton bound to that state. For acceptor states

$$\frac{E_{ex}^B}{E_A} \simeq 0.1$$

---

[2] This is also known as Haynes'Rule: J. R. Haynes: Phys. Rev. Lett. **4**, 361 (1960).

and for donors

$$\frac{E_{ex}^B}{E_D} \simeq 0.2$$

where E_{ex}^B is the binding energy of the exciton bound to the appropriate defect and E_A and E_D are the ionization energies of the localized acceptor and donor states, respectively. It should be noted that

$$E_{ex}^B = E_g - (hv)_{ex}^B - E_{ex}^{Free}$$

where E_{ex}^{Free} is the free exciton binding energy and $(E_{ex}^{Free} + E_{ex}^B)$ is the total ionization energy of the bound exciton. By use of these relationships, excitons found in the luminescence spectra can be associated with lines in the edge emission series.

4.2.3 Phonons

A characteristic of both exciton and edge emission spectra in II-VI compounds is that all emissions appear as a series of lines separated by equal energy increments. This is due to strong coupling to the lattice which results in more radiative processes being accompanied by phonon emission processes. Phonons are the quantized collective vibrational modes of the atoms comprising the crystal. Thus, the emission energies observed for excitons and the edge emission are

$$hv = E_g - E_i - nE_p; \quad n = 0, 1, 2, 3, \ldots$$

where E_i is the ionization energy of the exciton or localized state and E_p is the energy of the phonon involved. In most of the II-VI compounds, the most probable phonon to be emitted is the longitudinal optical (LO) phonon. For strong emission lines, it is not uncommon to find a transition appearing as a series of lines. This series consists of the no-phonon line followed by the transitions corresponding to the emission of $n = 1, 2, 3, \ldots$ phonons. The phonon coupling and the broadening by a high concentration of donor-acceptor states is responsible for the observed broad-band transitions.

Hopfield [4.66] has shown that the intensities of the phonon replicas generally can be described by the relation

$$I_n = I_0 \frac{\bar{N}^n}{n!}$$

where I_n is the relative intensity of the $n + 1$ lines of the set involving the emission of a photon plus n-LO phonons. $\bar{N}$, the mean number of emitted LO phonons given experimentally by the ratio I_1/I_0, provides a measure of the coupling of the center to the lattice.

4.2.4 Broad-Band Luminescence

Broad-band luminescence observed in ZnS and other II-VI compounds results from the doping of acceptor-type group IB elements (Cu, Ag, Au) (usually called "activators") as well as donor-type group IIIA (B, Al, Ga, In, Tl) or group VIIA (F, Cl, Br, I) elements (coactivators). These elements are associated with rather deep centers and produce various types of broad-band luminescence. In ZnS, for example, five types of luminescent transitions are observed when Cu is used as the activator: 1) Cu-green for nearly equal concentrations of Cu and a coactivator, 2) Cu-blue for higher concentrations of Cu than the coactivator, 3) Cu-red for Cu only, 4) Cu, (Ga, In)-red for relatively high and equal concentrations of Cu and Ga or an In coactivator, and 5) blue (self-activated, SA) for the coactivator only. The atomic structure of these luminescence centers as well as the nature of the luminescence transitions has been the subject of numerous investigations during the past two decades.

Polarization measurements [4.67] have shown that the Cu-red luminescence center is formed by the tight association of a substitutional Cu^+ ion with an S^{2-} vacancy at one of the nearest sites. Cu-green luminescence is known to arise from the donor-acceptor pair-emission mechanism, provided the donor and acceptor states involved are created by the coactivator and activator, respectively. This conclusion is drawn from the experimental observation of

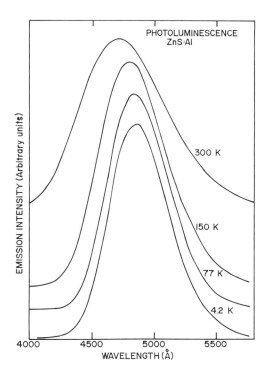

Fig. 4.6. Temperature dependence of photoluminescence spectra of Al-doped ZnS

the shift of emission peak to lower energies during decay and with decreasing excitation intensity [4.68] along with the fact that the decay curves obey the power-law relation and that the decay rate becomes faster with increasing photon energy in the emission band [4.69]. Concerning the Cu-blue luminescence, details of the electronic transition and the nature of the center are not so clear. The luminescence transition is though to be of the intra-center type which takes place between two localized centers, as inferred from the absence of time and intensity shifts [4.68].

Self-activated luminescence is produced by an electronic transition between localized states of a center which consists of either a group IIIA or a group VIIA impurity (coactivator) associated with a Zn vacancy center [4.70]. This model is supported by correlation of various experimental results such as the temperature dependence of the luminescence band shape and peak position [4.71], polarization of the luminescence [4.73], time-decay characteristics of the luminescence [4.73], and electron spin resonance [4.74]. The temperature dependence of luminescence spectra has been interpreted in terms of the simple configurational coordinate model [4.75]. One of the main features of the emission band is that as the temperature of the sample is increased from 4.2 K which, in turn, decreases the bandgap energy, the emission peak shifts toward higher energy and the emission band increases in width according to a sqare-root hyperbolic cotangent. This temperature-dependent luminescence characteristic is illustrated in Fig. 4.6 by the photoluminescence of an Al-doped ZnS single crystal.

4.3 Electroluminescence

As pointed out earlier, wide bandgap II-VI compounds are potentially efficient visible electroluminescent materials. It has also been pointed out that the main problem with these materials has been that due to self-compensation by native defects, it is difficult to obtain the *pn* junctions necessary for efficient injection electroluminescence. In order to circumvent this problem, many alternative approaches making use of metal-insulator-semiconductor (MIS) structures, heterojunction structures, *pn* junctions from solid solutions, and less complex metal-semiconductor structures have been attempted with some degree of success. With the recent progress in material and device technology, increased understanding and control of II-VI compounds have been acquired. As a result, more efficient visible LEDs can be prepared from II-VI compound *pn* junction diodes which has created renewed interest in this material.

Another reason for continued interest in II-VI compound injection luminescent devices is the search for deep-green-to-blue light-emitting materials. Among II-VI compounds, ZnSe and ZnS are good candidates for such applications. Other materials such as SiC and GaN having a comparable spectral range are not without problems. The major problem with GaN has been in achieving p-type doping, whereas the development of SiC diodes is complicated by crystal-growth conditions.

Another area of substantial progress involves ZnS-type [4.76 – 78] *DC* EL phosphors for applications such as flat-panel TV display [4.79]. These devices exhibit such properties as high efficiency, high contrast, high discrimination ratio, and an acceptable operating lifetime. Basic fabrication technology can result in inexpensive and rugged devices which may be constructed in unlimited symbol configurations and in a wide range of sizes. Typical *DC*-excited cells fabricated from ZnS:Mn, Cu powders [4.76] have an emission peak at 5800 Å. These devices operate typically at 100 V with a maximum current density of up to 5 mA/cm^2. Brightness of 30–100 fL and power efficiency of 0.1–0.3% have been obtained. These phosphors can be operated at a half-brightness lifetime of greater than 500 h. These phosphor systems, because of their potential advantages, are emerging as an important part of display technology.

4.3.1 Mechanism of Electroluminescence

In order to produce an efficient injection-electroluminescent device, the transport mechanism of the device must be consistent with the optical transition of the system. Several of the carrier-injection mechnisms for achieving radiative recombination are provided by minority-carrier injection, tunneling, and impact ionization. Device structures essential for such carrier injection are metal-semiconductor, metal-insulator-semiconductor, heterojunction, and *pn* junction. The most efficient method of achieving radiative recombination involves minority-carrier injection by *pn* junctions. In most of the II-VI compounds, the diffusion length of electrons is much greater than that of holes; therefore, electron injection into the *p*-side would dominate the device current (assuming that a *pn* junction can be produced in II-VI compounds) and the radiative recombination might be expected to occur in the *p*-type region. This implies that the length of the *p*-type region x_p must be >1 µm in order to obtain a high probability of recombination in the *p*-type region. In most of the previous efforts to achieve electroluminescence in II-VI compounds, it has generally been difficult to obtain the *pn* junctions required for efficient carrier injection. Other methods employed to achieve injection electroluminescence generally encounter a large number of interface states, and efficient carrier injection has been difficult to achieve. Therefore, in the case of the one-carrier system, a method of minority-carrier injection must be found which will yield high efficiency.

Optical transitions can be classified into two types—materials in which light emission results 1) via band-to-band or near band-to-band luminescent transitions and 2) via moderately deep centers ($\lesssim 10 \, kT$). In *pn* junction diodes, the high internal efficiency of the abundant band-to-band recombination sites can easily overcome the much smaller concentration of non-radiative recombination sites (killer centers), even under moderate operating conditions.

Therefore, the concentration of these killer centers—such as non-intentionally introduced foreign impurities, native defects, dislocations, and strains—must be minimized ($< 10^{16}$ cm^{-3}) in order to obtain fairly efficient radiative recombination. Furthermore, the strong absorption near the bandgap of II-VI compounds causes appreciable internal absorption of luminescence, and this self-absorption tends to be more severe in *p*-type or closely compensated *n*-type semiconductors because of the shift of the absorption edge to lower energies. Because of this strong self-absorption of the internally generated light, the external efficiency can be much smaller than the internal efficiency. This suggests that the optical path lengths in the *p*-type region should be minimized in order to obtain radiative recombination which is relatively free of self-absorption. Efficient luminescence in impurity states generally results from a free-carrier-to-un-ionized luminescence transition where the more likely event is a free-electron-to-bound-hole transition. The limiting factor in the device performance in such transitions is the competition between radiative and non-radiative sites which is due to the comparable concentrations of radiative and non-radiative centers. Therefore, for efficient injection electroluminescence, it is important to select the appropriate dopant species for the luminescent center and incorporate it into the device in the appropriate regions to optimum dopant concentrations.

4.3.2 Non *pn* Junction Structures: Experimental

Earlier efforts to achieve injection electroluminescence in II-VI compounds were concerned mainly with utilizing artificial barriers in the form of metal semiconductor (MS) [4.80] or metal-insulator semiconductor (MIS) by either deliberately providing thin insulating films between metal and semiconductor [4.81] or using a metal-alloy process [4.82]; these structures provide carrier injection into the semiconductor, which is necessary for radiative recombination, under both forward and reverse bias conditions. Diodes were also constructed from heterojunction configurations such as $Cu_2S - (ZnSe$ or $ZnS)$ structures [4.83] or ZnTe-CdS structures [4.84]. Other investigators found that light emission can also be obtained from the grain boundary [4.85] or as a result of bulk effects [4.86]. Although these devices provided substantial academic interest, efficiencies were generally low or operation at cryogenic temperatures (~ 77 K) was necessary.

The search for suitable barriers and greater understanding of the nature of the surface-barrier effects involved in metal-semiconductor and metal-insulator-semiconductor structures which will provide efficient injection electroluminescence in a given II-VI compound system involves problems of considerable complexity. Surface properties and barrier heights in II-VI compounds have been investigated in detail [4.87]. More recently, transport properties and the mechanism of electroluminescence in Schottky barrier structures have been examined carefully [4.88]. Along this line, ZnSe:Mn diodes having brightness of 500 fL and a yellow emission centered around 5785 Å have been

Table 4.6. II-VI compound electroluminescent MIS devices

System	Operation	Color	Peak wavelength	η_{ext} (%)	B/J fL/A/cm^2	Comments	Ref.
ZnTe:O	Au-i-(p-ZnTe)	Red	6850	0.3		Proton	4.91
			6580(77 K)	0.4(77 K)		bombarded	
iso-	6–80 V	Green	5580	0.02		Avalanche	
electronic	2–400 mA		5380(77 K)	0.1(77 K)		diode	
ZnTe:Li,P	Au-i-(p-ZnTe)	Yellow	5780	0.1–0.2		Al-vapor	4.92
ZnTe:Al		Yellow-green	5700(77 K)	20		diffusion	
D-A pair	25 mA/cm^2	Green	5470(77 K)	10		Avalanche	
	(average)					diode	
CdS		Red	6500			Collodion film	4.93
	6.7 V, 40 mA	Green	5150	$4 \cdot 10^{-4}$		hole injection	
	4 V, 40 mA		5052(77 K)	0.014		device	
ZnS:Al	Ag-ZnS						
V_{Zn}-Al$_{Zn}$	9 V, 0.3 A/cm^2	Blue	4600		130	Reverse bias	4.94
ZnS:Al	10 mA	Blue	4650	0.05		Forward bias	4.95
D-A pair				0.15(77 K)			

reported [4.89]. A metal-barrier diode produced from $ZnS_{0.6}Se_{0.4}$ mixed crystal showed green electroluminescence centered around 5450 Å with a brightness of 200 fL at a bias of 11 V [4.90]. Much progress is also being made in ZnS Schottky barrier structures where improvements in doping methods, ohmic contacts, and hole-injecting contacts are reported [4.6].

Progress has also been made in diodes having MIS structures (see summary in Table 4.6). In these devices, the methods used to provide insulating layers vary considerably. *Donnelly* et al. [4.91] obtained a semi-insulating ZnTe region several microns thick through the use of proton bombardment. External quantum efficiencies (η_{ext}) of 0.3% in the red (6850 Å) emission band and 0.02% in the green (5580 Å) were obtained in these devices. An insulating region was also produced in ZnTe samples by Al-vapor diffusion [4.92]. An external quantum efficiency of up to 0.2% was reported in the yellow (5780 Å) band at an average current density of 25 mA/cm^2. In CdS, MIS structures were obtained using collodion (cellulose dinitrate) or SiO_x as the insulator [4.93]. In these devices the threshold voltage of light emission was less than 1.5 V, with 12-mA diode current under forward-bias (hole-injection) conditions. Green emission peaking at 5150 Å was observed at room temperature. However, the external quantum efficiency was low, being on the order of 10^{-4}% under 6.7 V, 40-mA bias conditions. Blue emission was obtained from Al-doped ZnS MIS diodes, the emission peak centering around 4600 Å with brightness of 130 fL at 9 V and 0.3 A/cm^2 [4.94]. High-efficiency ZnS blue-light-emitting diodes with $\eta_{ext} = 0.05$% were also achieved recently [4.95].

Various injection electroluminescence devices have been prepared from heterojunction ZnSe-ZnTe [4.96–98], ZnSe-SnO$_2$ [4.99], and CdS-CuGaS$_2$ [4.100] structures. Although two different materials were used to construct the devices, an "artificial" *pn* junction can be obtained from such construction. Both green and blue electroluminescence were observed at 77 K, mostly as a result of D – A pair recombinations. However, room-temperature efficiencies of these devices were low, being dominated mostly by red-band luminescence. Such low room-temperature efficiency might be expected since the lattice mismatch between the two semiconductors produces a large number of interface states.

Much of the current interest in non *pn* junction LED work in II-VI compounds was triggered by the recent achievement of efficient ZnSe injection electroluminescent devices [4.101], and it is instructive to review the properties of this device. Such a review will also serve to illustrate the types of problems involved in the study of injection electroluminescent devices in II-VI compounds. Devices having external quantum efficiencies of up to 0.1% at room temperature and brightness levels of 200 fL at 10 V and 10 mA have been obtained. The diodes emitted light having a yellow-orange appearance peaking at 5900 Å at 300 K. Al-doped ZnSe low-resistivity (~ 0.01 Ω-cm) samples were used for the diode fabrication. Diodes of various metal-semiconductor combinations (Au-ZnSe-Ag, Au-ZnSe-In, Ag-ZnSe-In) were fabricated, typical diode dimensions being $0.1 \cdot 0.1 \cdot 0.02$ cm. Various electrical and optical studies were conducted on *I-V* characteristics, *C-V* characteristics, brightness, quantum efficiency, and spectral characteristics.

The *I-V* characteristics of all three devices (Au-ZnSe-In, Ag-ZnSe-In, and Au-ZnSe-Ag) were measured at room temperature. A typical log-log plot of the *I-V* characteristics of the diodes is shown in Fig. 4.7. The *I-V* characteristics obey a power-law relationship over a wide range of applied voltages and have the form $I \propto V^n$, where the exponent n has the value 1 and 2 in the low-current regions and 7–9 in the region with steeply increasing currents.

The curves exhibit characteristics similar to those of one-carrier space-charge-limited current in a semi-insulator with traps as predicted by *Lampert* [4.102]. At low voltages, the current is initially carried by thermally excited carriers contained in the insulator which shows ohmic characteristics. Assuming that the contribution of the diffusion current is negligible,

$$J = e\mu_n n \left(\frac{V}{L} \right) \qquad (\text{A/cm}^2),$$

where V is the applied voltage, μ_n is the mobility, n is the density of free carriers, and L is the length of the electrically active region. At higher current densities, where the density of injected carriers exceeds the thermal-equilibrium carrier density, the current will depart from the ohmic behavior. The current density of this region is expressed in terms of space-charge-limited current given by

$$J = \frac{9}{8} \varepsilon\mu_n \left(\frac{V^2}{L^3} \right) \theta \qquad (\text{A/cm}^2).$$

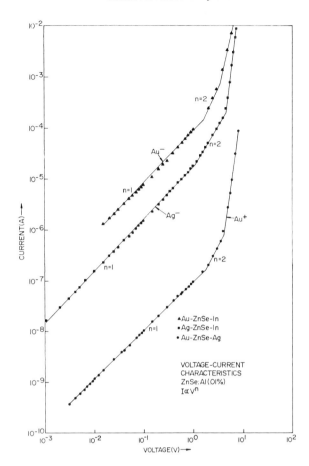

Fig. 4.7. Log-log plot of I-V characteristics of Ag-ZnSe-In, Au-ZnSe-In, and Au-ZnSe-Ag diodes at 300 K

Here θ is the ratio of free to trapped electrons and is given by

$$\theta = \frac{n}{n_t} = \frac{N_c}{N_t} \exp\left(-\frac{E_t}{kT}\right),$$

where N_c is the effective density of states in the conduction band, n_t is the density of traps, and E_t is the depth of traps from the bottom of the conduction band. A transition-region voltage V_{TR} at which the currents convert from ohmic to space-charge-limited behavior can be defined by equating the two current-density expressions. Thus,

$$V_{TR} \equiv (\tfrac{8}{9})\,(enL^2)/\varepsilon\theta \ \text{(volts)}.$$

The density of traps N_t can be determined from the trap-filled limit voltage V_{TFL} at which an abrupt increase in current occurs. Lampert has shown that at this limit,

$$N_t = \frac{1.1 \cdot 10^{-6} \varepsilon V_{TFL}}{L^2} \quad (\text{cm}^{-3}).$$

Values of trap depth E_t and density N_t for electron traps of 0.44 eV and $3 \cdot 10^{14}$ cm^{-3} are determined experimentally. These values are in good agreement with those determined by analysis of the thermally stimulated luminescence curves [4.103] of the Al-doped ZnSe samples.

In order to determine the transport mechanism in the steeply rising current region, the temperature dependence of the *I-V* characteristics must be measured. In the temperature range 80–300 K, the change in temperature did not produce much change in current value for a given voltage. The value of the current decreased with a decrease in temperature for a fixed value of voltage. For all temperature ranges, no significant change was seen in the slope of the *I-V* characteristics. These features strongly suggest that the dominant current-transport mechnism in this current range is tunneling. These diodes exhibited an inverse relation between the reverse-current knee and the doping concentrations of the bulk material as determined from the curve-tracer *I-V* characteristics. The reverse-current knee was taken to be the breakdown voltage. A general relation for the breakdown voltage in terms of the bulk doping concentration and the bandgap energy is given by [4.104]

$$V_B = 60 \left(\frac{E_g}{1.1}\right)^{3/2} \left(\frac{N_B}{10^{16}}\right)^{-3/4}.$$

The results of the calculation for various doping concentrations yield the upper limit for the breakdown voltage. For $N_B = 10^{18}$ cm^{-3}, the breakdown voltage is ~ 7.2 V and for $N_B = 5 \cdot 10^{17}$ cm^{-3}, the breakdown voltage is ~ 12.6 V. Both the temperature-dependent properties and the breakdown behavior indicate that the primary carrier-transport mechanism in these devices involves a mixture of tunneling and impact ionization.

The capacitance at zero applied bias allowed the determination of an effective depletion-layer width, which is created at the metal-semiconductor boundary. The depletion-layer width at the Ag and at the Au contact was found to be ~ 4 and 0.14 µm, respectively. When the Ag contact was biased negatively at a threshold voltage of 4 V, the field strength in the depletion layer was on the order of 10^4 V/cm. This field strength is sufficient to accelerate the electrons entering this region and to make them sufficiently "hot" to impact ionize luminescence centers.

Electroluminescence was observed at room, liquid-nitrogen, and liquid-helium temperatures. For these diodes the light-emission threshold as perceived by an unaided eye was ~ 4 V at room temperature. The room-temperature electroluminescence color changed from yellow to red as the doping concentration of the Al in the samples was increased from 10 to 1000 ppm. For a typical diode with Al concentration in the 100-ppm range, a yellow-orange band centered around 5900 Å at 300 K with a half-width of 800 Å is emitted, as shown in Fig. 4.8. When cooled to 4.2 K, this broad band was resolved into two bands peaking at 5600 Å and 6350 Å. Temperature-dependence studies of photoluminescence showed the 5900-Å yellow band observed at room temperature to be a self-activated luminescence band arising from transitions between localized states of a center consisting of a Zn vacancy and the impurity.

The semilog plot of brightness as a function of applied bias for a typical diode is given in Fig. 4.9. The room-temperature brightness value of 200 fL has been measured at 10 V, 10 mA. The value corresponds to the luminous efficiency of 200 fL/A/cm². The semilog plot of the square root of the voltage dependence upon brightness shown in Fig. 4.10 is linear at low voltages and sublinear at higher voltages. This dependence is related by a formula

$$B = B_0 \exp\left(-bV^{-1/2}\right),$$

where B is the brightness, V is the applied voltage, and B_0 and b are constants. The log-log of brightness as a function of device current is shown in Fig. 4.11. The brightness varies according to a brightness-current relationship of the

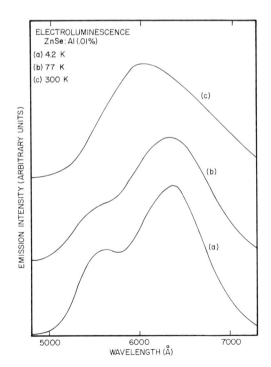

Fig. 4.8. Electroluminescence spectra of an Au-ZnSe-Ag diode at 4.2, 77, and 300 K

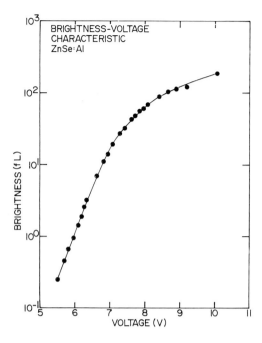

Fig. 4.9. Brightness-voltage characteristics of Au-ZnSe-Ag diode at 300 K

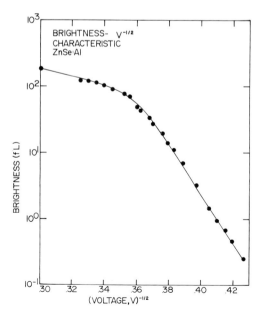

Fig. 4.10. Brightness vs. $V^{-1/2}$ characteristics of Au-ZnSe-Ag diode at
300 K

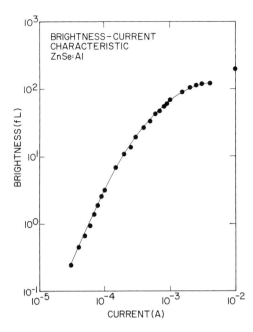

BRIGHTNESS–CURRENT
CHARACTERISTIC
ZnSe:Al

Fig. 4.11. Brightness-current relationship of Au-ZnSe-Ag diode at 300 K

form $B \propto I^n$, where the exponent n is ~ 2 in the low-current region and becomes
nearly linear in the high-current region. At higher current levels, the effect of
heating becomes apparent. The linear dependence of emission intensity of a
device current is often observed for radiative recombination associated with
a high-field impact-ionization process. The power dependence of current
$(B \propto I^n)$ and the exponential voltage relation $[B \propto \exp(-bV^{-1/2})]$ suggest that
the mechanism of the electroluminescence is due to impact ionization.

An external quantum efficiency of 0.5% has been obtained at 300 K for the
best diode fabricated. The measurement of the quantum efficiency as a function
of diode current is shown in Fig. 4.12 for a typical diode. Cooling the devices
to 4.2 K resulted in a factor of ten improvement in emission efficiency. It is
observed that the external quantum efficiency increases with increased doping
concentration. The optimum efficiency is obtained with 500-ppm Al-doped
samples. With further increase in doping concentration, the quantum efficiency
decreased. One can expect this increase in efficiency with an increase in number
of carriers for Al concentration < 500 ppm if it is assumed that the efficiency
loss due to non-radiative recombination remains essentially constant. The
decrease in efficiency above 10^{18} cm^{-3} Al doping concentration might be
expected based upon the fact that the effect of band tailing begins to predominate to the extent that non-radiative recombination processes (concentration quenching) account for a large part of the total current.

The experimental evidence—such as cathodic origin of light emission,
linear dependence of brightness on current, I-V characteristics, threshold

voltage for light emission at above bandgap voltage, electric-field strength in the depletion layer, and inverse dependence of breakdown voltages upon carrier concentration—suggests impact ionization by electron on the host lattice or luminescence centers via tunneling through the depletion layer assisted by filled traps in the thin insulating interface layer.

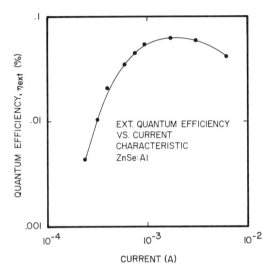

Fig. 4.12. External quantum efficiency vs current characteristics of Au-ZnSe-Ag diode at room temperature

4.3.3 *pn* Junctions: Experimental

In the previous sections it has been pointed out that the difficulty in achieving amphoteric doping in the production of *pn* junctions has been the major problem in the use of II-VI compounds, otherwise very efficient visible electroluminescent materials. Despite this difficulty, two types of *pn* junctions have been prepared from II-VI compounds previously—pseudo-binary junctions prepared from solid solutions and *pn* homojunctions prepared by ion implantation. Among the pseudo-binaries, $ZnSe_xTe_{1-x}$ and, more recently, ZnS_xSe_{1-x} structures have been of particular interest. Ion implantation has been used to produce various CdS, ZnTe, and ZnSe *pn* junction injection electroluminescent devices.

Efficient light emission has been reported in $ZnSe_xTe_{1-x}$ diodes [4.105] which exhibited external quantum efficiencies as high as 18% at 77 K with the emitted light peaking in the orange spectral region at 1.98 eV. However, the quantum efficiencies of such diodes at room temperatures was low—between 0.001 and 0.01%. The $ZnSe_xTe_{1-x}$ *pn* junction structure has been carefully examined by cathode-ray excitation [4.106]. The results show that the efficiency of the *n*-type region remained approximately the same throughout the temperature range 77–300 K, while that of the *p*-type region dropped by a factor of

10^2–10^3. Junction profiles obtained by cathodoluminescence, cathodoconductivity, and secondary electron emission indicated that the radiative recombinations occurred in the p-type region of the diode. The poor efficiency of the diode at room temperature originates from the low radiative recombination efficiency of the p-type $ZnSe_xTe_{1-x}$. In another study, stable, low-resistivity, p-type ZnS_xSe_{1-x} samples having a maximum hole concentration of $4 \cdot 10^{17}$ cm^{-3} were prepared by an unorthodox method, and pn junction diodes were made [4.18]. The diodes emitted light at ~ 2.1 eV with an estimated external quantum efficiency of $> 1\%$ at a bias voltage of 3.1 V and a diode current of 40 mA (40 A/cm^2). The group IIIA impurities (normally n-type dopants in II-VI compounds) were incorporated to produce p-type conductivity in ZnS_xSe_{1-x} materials. However, the nature of the shallow acceptor is not well understood at this time.

With the recent advent of ion-implantation technology, pn junction injection electroluminescent devices have been produced in II-VI compounds, and the results are encouraging. In earlier ion-implantation efforts, *Hou* et al. [4.45] reported observation of orange electroluminescence from F-ion implanted ZnTe diodes at -78 °C for a forward diode current of 15 mA (implanted samples were annealed at 550 °C for 2–5 h). Low-voltage electroluminescence was also observed at room temperature. The light intensity of the diode was proportional to the square of the current ($B \propto I^2$), which indicated that the recombination had taken place in the depletion region. In the Bi-ion implanted CdS, *Chernow* et al. [4.41] observed a green emission at 77 K from a forward bias with current pulses of 0.5 A. Electroluminescence was also reported in CdS by *Shiraki* et al. from N-ion implanted diodes [4.107]. Three groups of emission spectra were observed in the diodes at 100 K—dark red (8000 Å), yellow-orange (6200 Å), and green (5000 Å). The green emission is suggested to originate from the depletion region of the pn junction and, from a careful correlation of photoluminescence results, is attributed to the A-exciton associated with the LO-phonon. The yellow-orange band corresponds to the PL peak observed in the substrate. This band is thought to be a native defect center such as a Cd vacancy or an Ag impurity center. The broad red band (tail reaching up to 2 μm) which originates from the p-type region of the diode is correlated with the acceptor-like defect center formed as a result of N-ion implantation. The exponential natures of the emission intensity and current with voltage are quite similar to those of the usual injection luminescence observed from the pn junction diodes.

Type-conversion and injection-electroluminescent diodes have also been reported in ZnSe by *Park* and *Chung* from Li-ion implantation [4.47]. Electroluminescence was observed in both implanted diodes and MS diodes at room and nitrogen temperature. For an implanted-layer diode, the light-emission threshold as perceived by the unaided eye at 77 K was 2.4 V at 0.07 mA under forward bias. At room temperature, the threshold was increased to ~ 5 V at 3.5 mA. The diode emitted the green band centered around 5200 Å at 77 K. The threshold for an MS diode, on the other hand, was increased tenfold, and more efficient light emission was observed when the diode was reverse biased. The emission band for the MS diode was 6200 Å at 77 K. The electroluminescence band at 5200 Å observed at 77 K corresponds to the green photoluminescence band observed at 77 K from UV excitation. The green band shown in Fig. 4.5 seems to be characteristic of the Li-implanted and annealed crystals. This figure shows the photoluminescence spectra of the Li-implanted ZnSe crystals at 4.2 K after various annealing stages. As implanted, the crystal showed no green band but faint orange luminescence at ∼6200 Å which is known as the self-activated emission. In high-conductivity crystals this orange band always appears very strong. As the annealing temperature is increased from 150 °C, the green band appears and becomes dominant over the orange band at 300–400 °C. The green band also appears in Li-doped ZnSe grown by the vapor phase and from the melt. A similar green band was also observed by other workers in Li-doped ZnSe [4.108]. Sharp emission lines due to bound excitons [4.58] appearing at higher energies also show intensity variation as the annealing temperature is varied. In high-conductivity crystals, the I_2 line at 4432 Å (2.797 eV) associated with excitons bound to a neutral donor is more dominant than the I_1 line at 4455 Å (2.783 eV) associated with excitons bound to a neutral acceptor, and in some crystals the I_1 line is completely absent. At an annealing temperature as low as 150 °C, the I_1 line appears. At 400 °C the strength of the I_1 line is comparable to that of the I_2 line. These results indicate a correlation between the I_1 line and the appearance of the green band from the injection electroluminescence in the Li-ion implanted ZnSe *pn* junction diodes.

More recently, *pn* junctions were achieved in ZnSe by room-temperature Na-implantation after annealing at ∼500 °C [4.25]. As a result of the Na-implantation, a *p*-type layer having resistivity of 15 Ω-cm, effective hole mobility of 7 cm^2/V-s, and an effective bulk hole concentration of $6 \cdot 10^{17}$ cm^{-3} has been obtained by assuming a junction depth of 1 μm and a uniformly doped implanted layer. Electroluminescence was observed in both forward- and reverse-bias conditions. At 77 K a blue-green band peaking at 5100 Å was observed under forward bias. These devices produced only weak red-orange electroluminescence at room temperature.

More detailed studies of injection luminescence in *pn* junction diodes prepared by P-ion implanted ZnSe were reported by *Park* and *Shin* [4.48]. ZnSe *p*-type layers having room-temperature resistivity of 10 Ω-cm, hole mobility of 6 cm^2/V s, and a hole concentration of 10^{17} cm^{-3} resulted from the P-ion implantation [4.29]. The diode *I-V* characteristics, spectral characteristics, output power, and efficiencies were studied. The diodes produced exhibited excellent rectification characteristics, and electroluminescence was observed in both forward and reverse bias at room temperature. The light originating from the junction region was emitted uniformly.

The *I-V* characteristics of the diodes obeyed an ideal *pn* junction relation $I \propto \exp(eV/nkT)$, with the value of the coefficient n being 1.2–1.7 at the low forward-current level. Figure 4.13 shows the *I-V* characteristics of a typical diode at room temperature, as obtained on the diode curve tracer. The forward I-V characteristic is shown in Fig. 4.13a. The breakdown voltage of ~ 30 V was obtained from the reverse bias shown in Fig. 4.13b.

In forward bias the diode exhibited an electroluminescence spectrum very similar to the red photoluminescence spectrum, with a peak at 6300 Å (1.97 eV) at room temperature as shown in Fig. 4.14. This luminescence is characteristic for a phosphorus center in the ZnSe [4.28]. When the diode is biased in the reverse direction, it emits light peaking at 5900 Å (2.10 eV) as shown in Fig. 4.15. The electroluminescence band observed under reverse bias corresponds to the characteristic yellow band observed in the photoluminescence of Al-doped ZnSe from 3650 Å UV excitation. Analysis of data on the temperature dependence of the photoluminescence spectra shows the band to be self-activated luminescence arising from transitions between localized states of a center consisting of a Zn vacancy and the impurity.

The emitted power of the 6300 Å line in the forward bias increased linearly with increasing current at low current levels (<6 mA), as shown in Fig. 4.16. The sublinear power dependence of the current, the shift of the peak to longer wavelength, and the line-width broadening at higher current levels show the heat to be generated locally. The power dependence of the electroluminescence in the reverse-bias breakdown region is similar to that in the forward-bias case. In the forward direction a power level of 1 μW is obtained at ~ 6.5 mA input current and 5 V applied bias, with 0.01% external quantum efficiency.

The linear dependence of the brightness upon current suggests that the recombination mechanism is a result of the carriers injected across the junction, since $I \propto \exp(eV/kT)$ and $B \propto \exp(eV/kT)$ which gives $B \propto I$. In forward bias the predominant luminescence transition is a P-center in a p-type region. Electrons injected across the depletion layer recombine with available holes in the p-type region at the P-center. In the reverse-bias case, the center which consists of a Zn vacancy and Al is the dominant radiative transition. Here

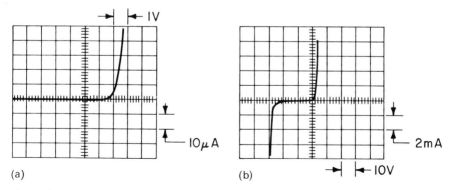

Fig. 4.13 a and b. Diode curve-tracer I-V characteristics of a P-implanted ZnSe diode at room temperature

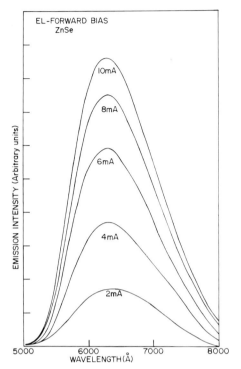

Fig. 4.14. Forward bias electroluminescence spectra of a phosphorus-implanted ZnSe diode at 300 K

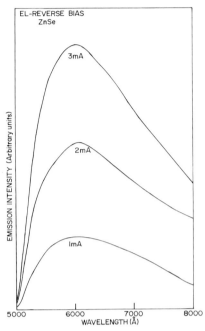

Fig. 4.15. Reverse bias electroluminescence spectra of a phosphorus-implanted ZnSe diode at 300 K

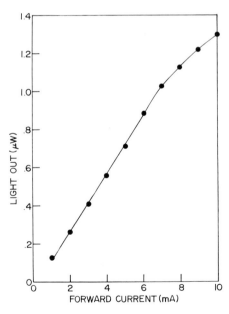

Fig. 4.16. Light output vs forward current of a phosphorus-implanted ZnSe diode

electrons may tunnel through the depletion layer to the n-type region, and as a result luminescence from the self-activated center is observed.

4.4 Future Prospects

For many years II-VI compounds have been known to be efficient visible luminescence materials. The realization of the potential of these materials as sources of LEDs, on the other hand, has long been hindered by the difficulty in forming the low-resistivity pn junctions essential for efficient injection electroluminescent devices. In recent years, with efforts to achieve efficient pn junctions, controlled doping has been performed on these materials by ion implantation with some degree of success. This has created renewed interest in II-VI compounds for further investigations on injection electroluminescence.

Two potential applications areas exist in II-VI DC EL phosphors and blue LEDs. The operational capabilities of the DC EL phosphors have already been demonstrated. For example, TV picture displays using ZnS:Mn electroluminescent thin-film devices have been fabricated, with device parameters of 40:1 contrast, 30-fL luminance operating at ~ 200 V. In blue LEDs, ZnSe and ZnS are considered to be potential candidates. Although in these materials some progress has been made in material preparation and device processing (such as formation of ohmic contacts), a proven technology is not available to date. In ZnSe it is desirable to obtain reproducible low-resistivity p- and n-type samples and pn junctions which emit light in the band-edge region. In addition, in ZnS doping conditions must be determined which involve relatively deep centers in order to obtain efficient room-temperature blue luminescence. In this regard, ion-implantation studies have indicated strong possibilities, and these investigations should be pursued. At the same time, the most important problem in II-VI compounds—methods of material preparation—must be re-examined carefully in order to obtain defect-free samples.

References

4.1   A. G. Fischer: in *Luminescence of Inorganic Solids*, ed. by P. Goldberg (Academic Press, New York 1966) Chap. 10
4.2   M. Aven: *II-VI Semiconducting Compounds*, ed. by D. G. Thomas (Benjamin Publ. Co., New York 1967) p. 1232
4.3   F. F. Morehead: *Physics and Chemistry of II-VI Compounds*, ed. by M. Aven, J. S. Prener (North-Holland Publ. Co., Amsterdam 1967) p. 611
4.4   A. G. Fischer: Solid-State Electron. **2**, 232 (1961)
4.5   A. G. Fischer: J. Electrochem. Soc. **118**, 139C (1971)
4.6   M. Aven, J. Z. Devine: J. Luminesc. **7**, 195 (1973)
4.7   D. J. Stukel, R. N. Euwema, T. C. Collins, F. Herman, R. L. Kortum: Phys. Rev. **179**, 740 (1969)
4.8   Y. S. Park, F. L. Chan: J. Appl. Phys. **36**, 800 (1965)

4.9   J. L. Birman: Phys. Rev. **115**, 1493 (1959)

4.10   D. C. Reynolds, S. J. Czyzak: Phys. Rev. **79**, 543 (1950)

4.11   L. C. Greene, D. C. Reynolds, S. J. Czyzak, W. M. Baker: J. Chem. Phys. **29**, 1375 (1958)

4.12   E. A. Davis, E. L. Lind: J. Phys. Chem. Sol. **29**, 79 (1968)

4.13   F. F. Morehead, G. Mandel: Phys. Rev. **137**, A924 (1965)

4.14   A. Catano, Z. K. Kun: J. Cryst. Growth **33**, 324 (1976)

4.15   S. Larach, R. E. Shrader, C. F. Stocker: Phys. Rev. **108**, 587 (1957)

4.16   F. F. Morehead, G. Mandel: Phys. Rev. **137**, A924 (1965)

4.17   M. Aven, W. Garwacki: Appl. Phys. Lett. **5**, 160 (1964)

4.18   R. J. Robinson, Z. K. Kun: Appl. Phys. Lett. **27**, 74 (1975)

4.19   Z. K. Kun, R. J. Robinson: J. Electron. Mat. **5**, 23 (1976)

4.20   G. Mandel: Phys. Rev. **134**, A1073 (1964)

4.21   H. H. Woodbury, M. Aven: In *Radiation Damage in Semiconductors* (Dounod, Paris 1965) p. 179;
        R. A. Anselmo, H. H. Woodbury: Bull. Am. Phys. Soc. **9**, 248 (1964)

4.22   W. Hayes: Phys. Rev. **138**, A1227 (1965)

4.23   Y. S. Park, P. M. Hemenger, C. H. Chung: Appl. Phys. Lett. **18**, 45 (1971)

4.24   J. L. Merz, K. Nassau, J. W. Shiever: Phys. Rev. B**8**, 1444 (1973)

4.25   A. J. Rosa, B. G. Streetman: J. Luminesc. **10**, 211 (1975)

4.26   G. B. Stringfellow, R. H. Bube: Phys. Rev. **171**, 903 (1968)

4.27   J. H. Haanstra, J. Dieleman: J. Electrochem. Soc. (Ext. Abstr.) **14**, 2 (1965)

4.28   A. R. Reinberg, W. C. Holton, M. de Wit, R. K. Watts: Phys. Rev. B**3**, 410 (1971)

4.29   Y. S. Park, B. K. Shin, D. C. Look, D. L. Downing: in *Ion Implantation in Semiconductors, Science and Technology* (Plenum Press, New York 1975) p. 245

4.30   A. K. C. Ho, K. C. Kao: Rad. Effects **17**, 137 (1973)

4.31   P. W. Yu, Y. S. Park: Appl. Phys. Lett. **22**, 345 (1973)

4.32   J. L. Merz, H. Kukimoto, K. Nassau, J. W. Shiever: Phys. Rev. B**6**, 545 (1972)

4.33   M. Aven: Bull. Am. Phys. Soc. **11**, 248 (1964)

4.34   J. C. Bouley, P. Blanconnier, A. Herman, Ph. Ged, P. Henoc, J. P. Noblanc: J. Appl. Phys. **46**, 3549 (1975)

4.35   M. Aven, B. Segall: Phys. Rev. **130**, 81 (1963)

4.36   H. H. Woodbury: J. Appl. Phys. **36**, 2287 (1965)

4.37   B. Tell: J. Appl. Phys. **42**, 2919 (1971)

4.38   M. Aven: J. Appl. Phys. **38**, 4421 (1967)

4.39   M. Aven, R. E. Halsted: Phys. Rev. **137**, A228 (1965)

4.40   J. P. Donnelly, A. G. Foyt, E. D. Hinkley, W. T. Lindley, J. O. Dimmock: Appl. Phys. Lett. **12**, 303 (1968)

4.41   F. Chernow, G. Eldridge, G. Ruse, L. Wahlin: Appl. Phys. Lett. **12**, 339 (1968)

4.42   W. W. Anderson, J. T. Mitchell: Appl. Phys. Lett. **12**, 334 (1968)

4.43   S. L. Hou, J. A. Marley, Jr.: Appl. Phys. Lett. **16**, 467 (1970)

4.44   Y. Shiraki, T. Shimada, K. F. Komatsubara: J. Appl. Phys. **43**, 710 (1972)

4.45   S. L. Hou, K. Beck, J. A. Marley, Jr.: Appl. Phys. Lett. **14**, 151 (1969)

4.46   J. Marine, H. Rodot: Appl. Phys. Lett. **17**, 352 (1970)

4.47   Y. S. Park, C. H. Chung: Appl. Phys. Lett. **18**, 99 (1971)

4.48   Y. S. Park, B. K. Shin: J. Appl. Phys. **45**, 1444 (1974)

4.49   M. Aven, H. H. Woodbury: Appl. Phys. Lett. **1**, 53 (1962)

4.50   G. H. Blount, M. W. Fisher, R. G. Morrison, R. H. Bube: J. Electrochem. Soc. **113**, 690 (1966)

4.51   S. Adachi, Y. Machi: J. Appl. Phys. Japan **14**, 1599 (1975)

4.52   R. K. Watts, W. C. Holton, M. deWit: Phys. Rev. B**3**, 404 (1971)

4.53   W. W. Anderson, R. M. Swanson: J. Appl. Phys. **42**, 5125 (1971)

4.54   C. H. Henry, K. Nassau, J. W. Shiever: Phys. Rev. **4**, B2453 (1971)

4.55   J. Dieleman, J. W. DeJong, T. Meijer: J. Chem. Phys. **45**, 3178 (1966)

4.56   See, for example, D. C. Reynolds: in *Optical Properties of Solids* (Plenum Press, New York 1969) p. 239

4.57   K. Colbow: Phys. Rev. **141**, 742 (1966)

4.58   P. J. Dean, J. L. Merz: Phys. Rev. **178**, 1310 (1969)
4.59   D. G. Thomas, R. Dingle, J. D. Cuthbert: in *II-VI Semiconducting Compounds*, ed. by
       D. G. Thomas (Benjamin, New York 1967) p. 863
4.60   See, fo example, D. C. Reynolds, C. W. Litton, Y. S. Park, T. C. Collins: J. Phys. Soc. Japan,
       21, Supplement, 143 (1966)
4.61   J. J. Hopfield, D. G. Thomas, M. Gershenzon: Phys. Rev. Lett. **10**, 162 (1963)
4.62   D. G. Thomas, M. Gershenzon, F. A. Trumbore: Phys. Rev. **133**, A 269 (1964)
4.63   D. G. Thomas, J. J. Hopfield, W. M. Augstyniak: Phys. Rev. **140**, A 202 (1965)
4.64   See, for example, S. Ida: J. Phys. Soc. Japan **25**, 177 (1968)
4.65   R. E. Halsted, M. Aven: Phys. Rev. Lett. **14**, 64 (1965)
4.66   J. J. Hopfield: J. Phys. Chem. Solids **10**, 110 (1959)
4.67   S. Shionoya, K. Urabe, T. Koda, K. Era, H. Fujiwara: J. Phys. Chem. Sol. **27**, 865 (1966)
4.68   K. Era, S. Shionoya, Y. Washizawa: J. Phys. Chem. Sol. **29**, 1827 (1968)
4.69   K. Era, S. Shionoya, Y. Washizawa, H. Ohmatsu: J. Phys. Chem. Sol. **29**, 1843 (1968)
4.70   T. Koda, S. Shionoya: Phys. Rev. **136**, A 541 (1964)
4.71   J. S. Prener, D. J. Weil: J. Electrochem. Soc. **106**, 409 (1959)
4.72   T. Koda, S. Shionoya: Phys. Rev. Lett. **11**, 77 (1963)
4.73   S. Shionoya, K. Era, H. Katayama: J. Phys. Chem. Sol. **26**, 697 (1965)
4.74   J. Schneider, A. Raüber, B. Dischler, T. L. Estle, W. C. Holton: J. Chem. Phys. **42**, 1839 (1965)
4.75   S. Shionoya, T. Koda, K. Era, H. Fujiwara: J. Phys. Soc. Japan **19**, 1157 (1964)
4.76   A. Vecht: J. Luminesc. **7**, 213 (1973)
4.77   J. M. Fikiet, J. L. Plumb: J. Electrochem. Soc. **120**, 1238 (1973)
4.78   T. N. Chin, L. A. Boyer: Solid-State Electron. **16**, 143 (1973)
4.79   H. Kawarada, N. Ohshima: Proc. IEEE **61**, 907 (1973)
4.80   A. G. Fischer: Phys. Lett. **12**, 313 (1964)
4.81   A. G. Fischer, H. I. Moss: J. Appl. Phys. **34**, 2112 (1963);
       R. C. Jacklevic, D. K. Donald, J. Lambe, W. C. Vassell: Appl. Phys. Lett. **2**, 7 (1963);
       M. G. Miksic, G. Mandel, F. F. Morehead, A. A. Onton, E. S. Schlig: Phys. Lett. **11**, 202
       (1964);
       D. D. O'Sullivan, E. C. Malarkey: Appl. Phys. Lett. **6**, 5 (1965);
       J. H. Yee, G. A. Condas: Solid-State Electron. **11**, 419 (1968)
4.82   B. L. Crowder, F. F. Morehead, P. R. Wagner: Appl. Phys. Lett. **8**, 148 (1966);
       D. I. Kennedy, M. J. Russ: J. Appl. Phys. **38**, 4387 (1967);
       D. P. Bortfeld, H. P. Kleinknecht: J. Appl. Phys. **39**, 6104 (1968)
4.83   M. Aven, D. A. Cusano: J. Appl. Phys. **35**, 606 (1964)
4.84   M. Aven, W. Garwacki: J. Electrochem. Soc. **110**, 401 (1963);
       T. Ota, K. Kobayashi, K. Takahashi: Solid-State Electron. **15**, 1387 (1972)
4.85   J. P. Donnelly, F. T. J. Smith: Solid-State Electron. **13**, 516 (1970);
       R. U. Khokhar, D. Haneman: J. Appl. Phys. **44**, 1231 (1973)
4.86   S. S. Yee: Solid-State Electron. **10**, 1015 (1967);
       A. N. Georgobiani, P. A. Todua: J. Luminesc. **5**, 14 (1972)
4.87   R. K. Swank: Phys. Rev. **153**, 844 (1967);
       C. A. Mead: Phys. Lett. **18**, 218 (1965);
       R. K. Swank, M. Aven, J. Z. Devine: J. Appl. Phys. **40**, 89 (1969);
       C. A. Mead: Solid-State Electron. **9**, 1023 (1966)
4.88   J. W. Allen: J. Luminesc. **7**, 228 (1973);
       A. W. Livingstone, K. Turvey, J. W. Allen: Solid-State Electron. **16**, 351 (1973);
       J. W. Allen, A. W. Livingstone, K. Turvey: Solid-State Electron. **15**, 1363 (1972)
4.89   M. E. Özsan, J. Woods: Solid-State Electron. **18**, 519 (1975)
4.90   M. E. Özsan, J. Woods: Appl. Phys. Lett. **25**, 489 (1974)
4.91   J. P. Donnelly, A. G. Foyt, W. T. Lindley, G. W. Iseler: Solid-State Electron. **13**, 755 (1970)
4.92   J. Gu, K. Tonomura, N. Yoshikawa, T. Sakaguchi: J. Appl. Phys. **44**, 4692 (1973)
4.93   D. J. Wheeler, D. Haneman: Solid-State Electron. **16**, 875 (1973)
4.94   Y. S. Park, B. K. Shin: unpublished
4.95   H. Katayama, S. Oda, H. Kukimoto: Appl. Phys. Lett. **27**, 697 (1975)

4.96 S. Fujita, S. Arai, K. Itoh, F. Moriai, T. Sakaguchi: Appl. Phys. Lett. **20**, 317 (1972);
      S. Fujita, S. Arai, F. Moriai, T. Sakaguchi: J. Appl. Phys. Japan **12**, 1094 (1973);
      S. Fujita, S. Arai, K. Itoh, T. Sakaguchi: J. Appl. Phys. **46**, 3070 (1975)
4.97 G. Le Floch, H. Arnould: Solid-State Electron. **16**, 941 (1973)
4.98 H. J. Lozykowski, H. L. Oczkowski, F. Firszt: J. Luminesc. **11**, 75 (1975)
4.99 K. Ikeda, K. Uchida, Y. Hamakawa, H. Kimura, H. Komiya, S. Ibuki: In *Luminescence of Crystals, Molecules, and Solutions*, ed. by F. Williams (Plenum Press, New York 1973) p. 245;
      K. Ikeda, K. Uchida, Y. Hamakawa: J. Phys. Chem. Solids **34**, 1985 (1973)
4.100 S. Wagner, J. L. Shay, B. Tell, H. M. Kasper: Appl. Phys. Lett. **22**, 351 (1973);
      S. Wagner: J. Appl. Phys. **45**, 246 (1974)
4.101 Y. S. Park, C. R. Geesner, B. K. Shin: Appl. Phys. Lett. **21**, 567 (1972)
4.102 M. A. Lampert: Phys. Rev. **103**, 1648 (1956)
4.103 E. T. Rodine: private communication
4.104 S. M. Sze, G. Gibbons: Appl. Phys. Lett. **8**, 111 (1966)
4.105 M. Aven: Appl. Phys. Lett. **7**, 146 (1965)
4.106 M. Aven, J. Z. Devine, R. B. Bolon, G. W. Ludwig: J. Appl. Phys. **43**, 4136 (1972)
4.107 Y. Shiraki, T. Shimada, K. F. Komatsubara: In *Ion Implantation in Semiconductors and Other Materials*, ed. by B. L. Crowder (Plenum Press, New York 1973) p. 395
4.108 S. Ibuki, H. Koniya, A. Mitsuishi, A. Manabe, H. Yoshinaga: In *II-VI Semiconducting Compounds*, ed. by D. G. Thomas (Benjamin, New York 1967) p. 1140

5. Chalcopyrites

S. Wagner

With 10 Figures

By 1973, when the first reports on light-emitting diodes based on chalcopyrite-type semiconductors were published, the key physical properties of numerous I-III-VI$_2$ and II-IV-V$_2$ compounds had become known. These included the characteristics indispensable for the evaluation of a potential electroluminescent material, such as magnitude and nature of the lowest bandgap, and electrical conductivity type. Since then the chalcopyrites CdSnP$_2$, CuGaS$_2$, CuInS$_2$, and CuInSe$_2$ have been incorporated in light-emitting diodes. This chapter is restricted to these four compounds. The reader who requires more extensive information about chalcopyrite-type semiconductors is referred to the books, conference records, and bibliographies listed in the reference section [5.1–9]. The recent book by *Shay* and *Wernick* [5.9] affords an excellent overview of both materials and physics aspects of chalcopyrites.

5.1 Properties of Chalcopyrites

5.1.1 Crystal Structure

Chalcopyrite-type semiconductors are ternary analogs of the sphalerite (zincblende) modifications of binary semiconductors. Ordered substitution of the metal in III-V or in II-VI compounds by two metals (II and IV, or I and III, respectively) doubles the identity period of the initially cubic unit cell. By definition, this doubling takes place along the z-direction resulting in a tetragonal superstructure of zincblende with $c/a = 2$ (Fig. 5.1). The prototype of this group of compounds is the mineral chalcopyrite, CuFeS$_2$, with structure type El$_1$ and space group I$\bar{4}$2d.

Many chalcopyrite properties are quasi-cubic in the sense of being derived from those of the analogous zincblende compounds. However, the noncubic aspects, being unique for chalcopyrites, command substantial interest. The three principal structural noncubic features are the doubled unit cell, the tetragonal distortion, and the displacement of the group V or VI atoms.

The tetragonal distortion, usually a compression and rarely an elongation of the unit cell along the c-axis, is quantified as $2 - c/a$. For tetragonal compression, $2 - c/a$ is positive. From Table 5.1 it can be seen that the unit cells of CdSnP$_2$ and CuGaS$_2$ are clearly compressed and that the cells of CuInS$_2$ and CuInSe$_2$ are slightly elongated if not virtually quasi-cubic. This last remark is in order because different investigators find differing values for the unit cell parameters. These discrepancies arise from the study of crystals with different composition. A well documented case is the dependence of the a-parameter of $CuGaS_2$ on the concentration of Ga_2S_3 [5.10] and the large drop of a in going

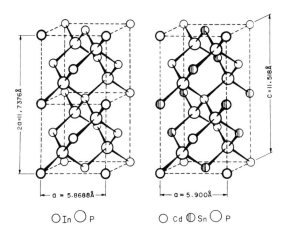

Fig. 5.1. The unit cells of zinc-blende (sphalerite) InP, and of chalcopyrite $CdSnP_2$. Note the ordered substitution of In by Cd and Sn. From [5.87]

Table 5.1. Structural and electronic properties of chalcopyrites

	$CdSnP_2$	$CuGaS_2$	$CuInS_2$	$CuInSe_2$
$a(\text{Å})$	5.900	5.33 to 5.35	5.523	5.782
$c(\text{Å})$	11.513	10.47	11.133	11.620
c/a	1.951[a]	1.96 to 1.97[h]	2.016[l]	2.010[p]
x (free parameter)	0.265[b]	0.275[i]	0.214[i]	0.224[i]
melting point (°C)	565 (perit.)[c]	1240[h]	>1050[m]	986[p]
$\Delta H_{f,298}^0(\text{kJmol}^{-1})$	67 (calc.)[d]			
$E_g(300\,\text{K})$	1.17[e]	2.40[j]	1.53[j]	1.01[q]
$N_D - N_A(\text{cm}^{-3})$	$\leq 5 \cdot 10^{18\,f}$	—	$\leq 3.5 \cdot 10^{17\,n}$	$\leq 4 \cdot 10^{17\,o}$
$\rho_n(\Omega\text{cm})$	$\geq 0.01^f$	—	$\geq 0.24^n$	$\geq 0.05^o$
$\mu_e(\text{cm}^2\text{V}^{-1}\text{s}^{-1})$	$\leq 2000^f$	—	$\leq 200^o$	$\leq 320^o$
m_e/m_0	0.036[g]			0.15 (est.)[r]
$N_A - N_D(\text{cm}^{-3})$		$\leq 2 \cdot 10^{18\,k}$	$\leq 1 \cdot 10^{17\,o}$	$\leq 1 \cdot 10^{18\,o}$
$\rho_p(\Omega\text{cm})$	$\geq 100^f$	$\geq 0.35^k$	$\geq 5^o$	$\geq 0.5^o$
$\mu_h(\text{cm}^2\text{V}^{-1}\text{s}^{-1})$		$\leq 20^k$	$\leq 20^n$	$\leq 10^o$
m_h/m_0		0.69[k]	1.3[n]	0.6 (est.)[r]

References to Table 5.1:

[a]   [Ref. 5.9, p. 7]   [g]   [Ref. 5.5, p. 288]   [m]   [5.56]

[b]   [Ref. 5.5, p. 52]   [h]   [5.10]   [n]   [5.31]

[c]   [5.28]   [i]   [5.73]   [o]   [5.25]

[d]   [Ref. 5.5, p. 174]   [j]   [5.103]   [p]   [5.32]

[e]   [5.98]   [k]   [5.30]   [q]   [5.105]

[f]   [5.85]   [l]   [Ref. 5.9, p. 4]   [r]   [5.79]

from the so-called green ($a=5.35$ Å) to the orange ($a=5.33$ Å) varieties of $CuGaS_2$ [5.11]. Indeed we encounter here one of the most pressing problems in chalcopyrite research, the characterization of compound stoichiometry and its effects on physical and chemical properties.

Displacement of the V or VI atoms is the principal noncubic aspect of atomic positions within the unit cell. These positions for A(I or II), B(III or IV) and C(VI or V) for the lower half of the unit cell are:

A in 000 and $0 \frac{1}{2} \frac{1}{4}$;

B in $00 \frac{1}{2}$ and $0 \frac{1}{2} \frac{3}{4}$;

C in $x \frac{1}{4} \frac{1}{8}$, $\bar{x} \frac{3}{4} \frac{1}{8}$, $\frac{3}{4} x \frac{7}{8}$ and $\frac{1}{4} \bar{x} \frac{7}{8}$.

(Positions in the upper half are reached by adding $\frac{1}{2} \frac{1}{2} \frac{1}{2}$). A and B positions are diamond-like ("cubic"), but the C positions are noncubic if $x \neq 0.25$. X-ray investigations of II-IV-V$_2$ compounds have shown that the four V ligands of the IV atom form a perfect tetrahedron. In that case the x-parameter is a function of, and can be predicted from, the tetragonal distortion [5.12], $x = 0.5 - (c^2/32a^2 - 1/16)^{1/2}$. A more complicated equation involving a bond deformation parameter has also been proposed [Ref. 2.5, p. 63]. By and large, though, the simple correlation given above holds and demonstrates that only one free noncubic parameter suffices for structural characterization of II-IV-V$_2$ chalcopyrites. Attempts to correlate the free x-parameters with c/a of I-III-VI$_2$ compounds have been less successful, implying that the tetrahedron formed around the group III atom is not perfect. It is easy to see that for $CuInS_2$ and $CuInSe_2$ with $c/a \cong 2.0$, x should be 0.25 when computed from the above equation. However, as can be seen in Table 5.1 it is substantially less. A very weak (110) X-ray reflex observed in $CuInSe_2$ implies that all three Se parameters may be noncubic [5.13].

The existence of two noncubic structural parameters is but one manifestation of the less cubic character of I-III-VI$_2$ chalcopyrites. This behavior results from significant participation of Group I element d-electrons in the interatomic bonding. d-levels are admixed to the uppermost valence band.

5.1.2 Band Structure

The most simple representation of the chalcopyrite band structure is obtained from a comparison of the exactly quasi-cubic tetragonal superlattice to the sphalerite cell [5.14]. Since the zincblende "primitive cell" (containing one formula unit) fills a quarter of the volume of the chalcopyrite "primitive cell" (with two formula units), the chalcopyrite Brillouin zone is four times smaller than that of zincblende. Therefore, each point of the chalcopyrite Brillouin zone corresponds to four equivalent points in the sphalerite Brillouin zone. For instance, the following energy levels of the zincblende Brillouin zone map into the $\Gamma(0,0,0)$ level of the chalcopyrite Brillouin zone: $\Gamma(0,0,0)$, $X(0,0,2\pi/a)$, $W(0,2\pi/a,\pi/a)$, $W(2\pi/a,0,\pi/a)$. One interesting consequence is the appearance of pseudo-direct transitions [5.15,16]. These are indirect transitions (e.g., $\Gamma_{15} \rightarrow X_1$) in analog zincblende compounds which, by zone folding, become direct (e.g., $\Gamma_{15} \rightarrow \Gamma_1$) in chalcopyrites. In an exactly quasi-cubic chalcopyrite the nature of these transitions is not altered. However, an initially forbidden transition will become allowed to the extent by which the crystal is perturbed by tetragonal distortion.

Under the combined influence of crystal field and spin-orbit splitting the original threefold degeneracy of the p-like valence band is lifted. For II-IV-V$_2$ compounds good agreement with experimental results is obtained in a quasi-cubic model where bandgap, and spin-orbit splitting Δ_{so}, are set equal to those of the binary analog (InP for CdSnP$_2$), and the crystal-field splitting Δ_{cf} is produced by the tetragonal distortion of the chalcopyrite unit cell [5.17]. The crystal-field splitting is well approximated by the linear relation $\Delta_{cf} = -b[1 - c/2a]$, with the deformation potential b typically 1 eV.

This quasi-cubic model does not agree with experimental results for I-III-VI$_2$ chalcopyrites. Bandgap energies are about 1 eV less than those of corresponding binary compounds (ZnS for GuGaS$_2$, Zn$_{0.5}$Cd$_{0.5}$S for CuInS$_2$, and Zn$_{0.5}$Cd$_{0.5}$Se for CuInSe$_2$); the spin-orbit splitting is also lower or even negative. This shortcoming of the quasi-cubic model results from neglect of the hybridization of anion p-levels and group I d-levels, which has been observed in electroreflectance [5.18] and X-ray photoemission [5.19–21], and which can be inferred from chemical shifts observed in NMR [5.22]. The amount of hybridization (d-like character) $1 - \alpha$, of the uppermost valence band has been obtained [5.18] from Δ_{so} observed in ternary compounds, Δ_p, the spin-orbit splitting of the p-like binary analog, and Δ_d, the (negative) spin-orbit splitting of the d-levels:

$$\Delta_{so} = \alpha \Delta_p + (1 - \alpha)\Delta_d.$$

Typically, the d-like character is $\sim 40\%$ for Cu-III-VI$_2$, and $\sim 20\%$ for Ag-III-VI$_2$ chalcopyrites. In a model containing two parameters, the energy separation between the (non-interacting) p- and d-bands, and the interaction strength between these bands, the reduction in bandgap and the magnitude of the spin-orbit splitting can be explained [5.23]. No adequate theory exists for the crystal field splitting which in I-III-VI$_2$ chalcopyrites is larger than the value estimated from the tetragonal compression formula given above.

All four compounds of interest in this chapter have direct gaps with valence band maxima and conduction band minima at Γ. The bottom of the conduction band Γ_1 is similar to Γ_1 in sphalerite compounds, with little anisotropy in effective masses of electrons. Because of the higher degree of ionicity of

I-III-VI$_2$ as compared to II-IV-V$_2$ compounds, the permitted valence zones of I-III-VI$_2$ are narrower and separated by wider forbidden sections than those of the II-IV-V$_2$ compounds. The apex of I-III-VI$_2$ valence zones is flatter than that of II-IV-V$_2$. Therefore, hole mobilities are lower in I-III-VI$_2$ than in II-IV-V$_2$ chalcopyrites [5.24].

A diagram of the lowest energy gaps of I-III-VI$_2$ and II-IV-V$_2$ compounds is presented in Fig. 5.2.

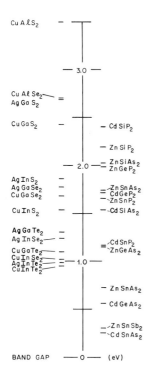

Fig. 5.2. Magnitude of the lowest energy gap of I-III-VI$_2$ and II-IV-V$_2$ compounds

5.1.3 Electrical Properties

In surveying chalcopyrites for *dc* electroluminescence applications, electrical properties are criteria second in importance only to the magnitude of the energy gap. For instance, CuAlS$_2$ is an attractive candidate for blue light emitting diodes. CuAlS$_2$ has band gap of 3.6 eV but it can be obtained only *p*-type. Exploitation of this compound will require a heterodiode with a matching *n*-type partner.

Minimum resistivities, maximum carrier concentrations, and Hall mobilities are listed in Table 5.1. The principal electrically active defects appear to be native, and are thought to be introduced by nonstoichiometry. They have not yet been conclusively identified. The Cu-vacancy was speculated to be the acceptor in Cu-III-VI$_2$ compounds [5.25]. In a recent EPR study [5.26] the copper vacancy with a trapped hole was positively identified in CuGaS$_2$. The hole is shared by the four sulfur neighbors, so that the acceptor actually is the complex $[V_{Cu}S_4]^{7-}$.

Whereas the electrical properties of II-IV-V$_2$ chalcopyrites are fixed by initial composition, doping, and conditions of crystal growth, the conductivity and even the conductivity type of numerous I-III-VI$_2$ chalcopyrites can easily be changed by annealing. Heating in chalcogen vapor produces p-type or high resistivity n-type crystals, while annealing in vacuum leads to high resistivity p-type, or n-type material. Anneals in saturated chalcogen vapor, i.e., in the presence of unevaporated chalcogen liquid, must be carried out so that the crystal is not in contact with liquid sulfur, selenium or tellurium. If it is, the crystal surface can be severely attacked and electrical properties become erratic, apparently by preferential dissolution of one of the component metals.

In II-IV-V$_2$ compounds a variation in the ratio of the constituent elements has been shown to lead to a wide range of resistivities in resulting crystals. However, in CdSnP$_2$ conductivity control is best achieved by doping. Crystals grown from Sn solution are low resistivity (~ 0.01 Ωcm) n-type. Cu-doping [5.27, 28] leads to high-resistivity ($\gtrsim 10$ Ωcm) p-type, Li and Ag-doping [5.28] to high resistivity n-type material.

Extensive electrical characterizations have been published on melt-grown and vapor-transported CuGaS$_2$ crystals annealed at maximum sulfur pressure between 250 and 850° [5.29, 30]. The two studies considerably differ in detail, presumably because of lack of precise control over the crystal composition. The main conclusions, though, are similar. CuGaS$_2$ can be made only p-type. Room temperature hole densities p were found to increase, and resistivities ρ to decrease with rising annealing temperature. p and ρ ranged from $4 \cdot 10^{15}$ cm^{-3} and 125 Ωcm to $2 \cdot 10^{18}$ cm^{-3} and 0.35 Ωcm. Room temperature mobilities μ lay between 5 and 20 cm^{2}V^{-1}s^{-1}. Low temperature Hall measurements [5.31] show that the crystals can be highly compensated with donor to acceptor ratios, N_D/N_A, ranging from 0.84 in as-grown highly resistive material to 0.11 in a highly conducting crystal obtained by sulfur anneal at 660 °C. Acceptor ionization energies ε_A range from 5 to 60 meV [5.29, Fig. 5.3], and from 73 to 390 meV [5.30], with low values for conducting and high values for resistive samples. The acceptor activation energy ε_A^o in the dilute limit was calculated to be 130 meV [5.30]. Larger experimental values are ascribed to effects of heavy compensation. From the dependence of the hole density p [5.29] or acceptor concentration N_A [5.30] on annealing temperature activation energies for the introduction of acceptors of 0.55 eV and 0.68 eV, respectively, were obtained. These values are ascribed to essentially the formation energy of a copper vacancy. Attempts to dope CuGaS$_2$ with likely foreign acceptors [5.29] such as group VIII transition metals (Ni, Pd), group IIB metals (Zn, Mg), or group VA elements (N, P) were unsuccessful although the occurrence of p-type conductivity in nitrogen- and phosphorus-doped samples was not ruled out.

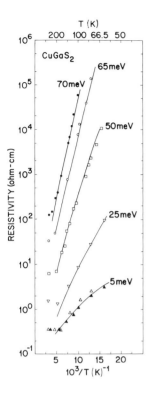

Fig. 5.3. Temperature dependence of the electrical resistivity of CuGaS$_2$ samples with different acceptor concentrations. From [5.29]

The effect of crystal growth conditions on electrical properties is demonstrated in a recent study on CuInS$_2$ [5.31]. This chalcopyrite can be obtained n- and p-type. Crystals grown from stoichiometric melts which had been homogenized at 1 080, 1 125, or 1 175 °C were n-type ($5.5 \cdot 10^{12}$ cm^{-3}), n-type ($5.4 \cdot 10^{16}$ cm^{-3}), and p-type ($1 \cdot 10^{17}$ cm^{-3}), respectively. Annealing in excess sulfur at 550 °C gave p-type samples with $\rho \cong 1$ Ωcm and $\mu \cong 20$ cm^{2}V^{-1}s^{-1}, while 725 to 800 °C anneals with In metal (as a sulfur sink) in the presence of CuInS$_2$ powder resulted in n-type material with $\rho \cong 1$ Ωcm and $\mu_n = 100$ to 200 cm^{2}V^{-1}s^{-1}. One shallow and one deep (350 meV) donor, and one acceptor (50 meV) level were identified. As in CuGaS$_2$, evidence was obtained for a high degree of compensation between native donors and acceptors.

Since a number of devices based on CuInSe$_2$ has been produced the electrical properties of this chalcopyrite have received much attention. Available in n- and p-type, with a comparatively low melting point of 986 °C, and containing a chalcogen with relatively low vapor pressure, CuInSe$_2$ is also a material more tractable than the high-melting sulfides. Boules obtained by slow cooling of exactly stoichiometric melts contain n- and p-type regions. When this material is zone leveled, uniformly n-type ($2.4 \cdot 10^{16}$ cm^{-3}) crystals result [5.32]. Crystals obtained by cooling of melts are n-type when the melt is slightly In rich, and p-type when slightly Se rich [5.33]. Annealing in saturated Se vapor makes

CuInSe$_2$ p-type while anneals under minimum Se pressure (in an evacuated ampoule in presence of powdered CuInSe$_2$) leads to n-type conductivity [5.25,32,33]. Cu [5.34], Zn [5.33,35], Cd [5.33,36,37], In [5.32,38], and Cl and Br [5.35] have been reported to act as n-type dopants when diffused [5.32–34, 36,38] or ion implanted [5.35,37] into p-type material. Since the anneals employed for impurity diffusion or for activation of ion implants by themselves can convert CuInSe$_2$ from p- to n-type, the electrical activity of some of these impurities has not been definitely established. A careful study of bulk-doped material seems to be in order.

EPR studies of Fe [5.39–41], Ni [5.41–43], and Mn [5.44] in CuGaS$_2$ and Mössbauer spectroscopy of Fe in CuGaS$_2$ [5.45] underscore the potential effect of transition metal impurities on the electrical characteristics of a chalcopyrite.

5.2 Technology of Materials and Devices

5.2.1 Phase Diagrams

No chalcopyrite is in equilibrium exclusively with its elemental components. Constitution diagrams are rendered complex by coexistent binary and even ternary phases, by polymorphism, and by wide existence ranges. Nevertheless numerous important equilibria have been investigated by judicious selection of quasi-binary diagrams, connecting an element and a binary compound (Sn-CdP$_2$) or two binary compounds (Cu$_2$S-Ga$_2$S$_3$), which contain the chalcopyrite of interest.

Partial Cd-CdSnP$_2$, Sn-CdSnP$_2$, and eutectic Cd + Sn-CdSnP$_2$ [5.46], and the entire Sn-CdP$_2$ quasi-binary [5.28] were studied in connection with the growth of CdSnP$_2$ crystals from dilute solutions in the constituent metals. CdSnP$_2$ is the only ternary compound in the Sn-CdP$_2$ quasi-binary. It is stoichiometric with a narrow existence range; it is in equilibrium with liquid Sn and orthorhombic α-CdP$_2$ up to $\sim 420\,^\circ$C, with the liquid Sn and tetragonal β-CdP$_2$ between $\sim 420\,^\circ$C and the peritectic at $\sim 570\,^\circ$C. Heated above $\sim 570\,^\circ$C, CdSnP$_2$ decomposes to β-CdP$_2$ and liquid Sn.

A recent extensive study of the CuGaS$_2$ equilibria [5.10] has ascertained the extensive existence range, $0 < y < 0.1645$ of Cu$_{1-y}$Ga$_{1+y/3}$S$_2$ in agreement with earlier work [5.47,48]. CuGaS$_2$ can also dissolve excess sulfur (Fig. 5.4). The melting point of exactly stoichiometric CuGaS$_2$ is 1 240 $^\circ$C, whereas the maximum melting (1 247 $^\circ$C) composition is obtained from a melt of 22 a/o (atomic percent) Cu, 25 a/o Ga, and 53 a/o S. This study has also shed light on the controversy about "dark", or "green", and "orange" CuGaS$_2$ [5.39,45, 48–51] which to the more discriminating investigator appear as the "black", "yellow", and "red" forms [5.52]. Efficient photo- and electroluminescence are observed in green, but not in orange crystals. The colorations are combined effects of variation in composition, and of contamination with iron. Un-

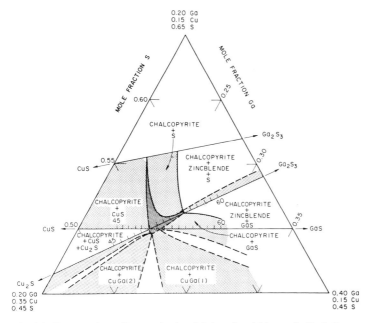

Fig. 5.4. Ternary phase diagram in the vicinity of stoichiometric CuGaS$_2$ at room temperature. The chalcopyrite phase field is cross-hatched. From [5.10]

contaminated CuGaS$_2$ is orange when Ga-rich, and green when grown with excess Cu. The green color arises from light scattering by Cu$_y$S inclusions [5.10,53]. The matrix of this green material is exactly stoichiometric and, with the ensuing maximum structural perfection, it is most efficient in luminescence. This pure CuGaS$_2$, green or orange, exhibits various hues of yellow and red when contaminated with Fe^{2+}, due to weak absorption in the green. When Fe^{2+} is oxidized to Fe^{3+} by anneals in sulfur, very strong absorption occurs in bands centered at 1.2 and 1.9 eV, and the crystal appears black [5.54]. This absorption is thought to arise from charge transfer between neighboring Cu$^+$-Fe^{3+} pairs [5.39] or the $3p$ orbitals of sulfur and the vacant $3d$ orbital of Fe^{3+} [5.55].

The only available phase information about CuInS$_2$ is a range of solidification temperatures of 1 000–1 050 °C [5.56].

A Cu$_2$Se-In$_2$Se$_3$ quasi-binary diagrams has been determined [5.57] and confirmed in the course of crystal growth studies [5.32]. Similar to CuGaS$_2$, CuInSe$_2$ exists in a range from strictly stoichiometric to In-rich compositions. At 810 °C the tetragonal chalcopyrite transforms to a zincblende phase due to disordering of the metal atoms. The zincblende modification melts at 986 °C.

Miscibility and alloy formation between chalcopyrites themselves, and between chalcopyrites and binary analogs are frequently extensive if not complete [Ref. 5.5, p. 194, 5.58–62].

5.2.2 Growth of Single Crystals

Chalcopyrite single crystals have been grown from near-stoichiometric melts, from solutions in liquid metals, or by chemical vapor transport. The starting materials are sealed in quartz ampoules to prevent volatilization of constituents and to exclude air. The ampoules are evacuated or backfilled with a low pressure of inert gas. Chemically inert liners, such as a boat or crucible of graphite or vitreous carbon, or a carbon film deposited on the inside wall of the quartz ampoule, may contain the melt or solution.

High pressures can develop during growth of phosphides or sulfides. To avoid these, binary compounds containing P, or S, are used as starting materials. Use of the pure elements requires slow heating to remove free P or S by reaction at low temperature. The formation of chalcopyrites is exothermic. This is another reason for inducing reaction at relatively low temperature to avoid sudden heating and buildup of excessive pressures.

$CdSnP_2$, which grows peritectically from β-CdP_2 and liquid Sn, has been crystallized from solutions in liquid Cd or Sn [5.63–67]. $10 \cdot 2 \cdot 0.5$ mm^3 skeletal crystals are obtained from 1 to 10 m/o $CdSnP_2$ solutions in Sn prepared by adding CdP_2 powder to Sn. Excess Sn is removed by dissolution in Hg, and Hg with dilute nitric acid and vacuum heating [5.67]. 0.1 to 1.5 mm thick platelets with up 0.5 cm^2 large faces are obtained from 10 m/o $CdSnP_2$ in Cd solutions. After crystal growth, the Cd can be distilled off in vacuum, not below 10^{-3} torr, at 350 °C [Ref. 5.5, p. 118].

$CuGaS_2$ and $CuInS_2$ crystals have been grown from near-stoichiometric melts, from solutions, and by chemical vapor transport.

Crystallization from near-stoichiometric $CuGaS_2$ melts at temperatures below 1 200 °C in silica boats contained in evacuated silica ampoules resulted in clearly separated clusters of orange and dark crystallites with dimensions of ~1 mm [5.49]. Cooling stoichiometric compositions from 1 140 °C led to dark crystals only. Stoichiometric mixtures cooled from 1 180 to 1 100 °C at a rate of 2 °C/h resulted in yellow-green to red $CuGaS_2$ crystals [5.30]. Poly-crystalline $CuGaS_2$ powders reacted from Ga_2S_3, Cu, and S at temperatures up to 1180 °C were sintered [5.68] below the melting point, then heated in silica ampoules above the melting point and cooled at 3 °C/h in a temperature gradient of 3 °C/cm [5.10]. Melting points ranged from 1 240 °C in the first to 1 200 °C in the last part to freeze, indicating a continuous change in composition when starting with stoichiometric melts. All melt compositions with excess sulfur or excess gallium resulted in orange crystals. Cooling of 10 m/o $CuGaS_2$ solutions in In, prepared from the elements, in evacuated quartz ampoules led to yellow-orange crystals [5.69]. Melts were cooled from 1 150 °C at a rate of 4 °C/h. Excess In was dissolved in hot HCl to isolate 10 mm long needles, or $5 \cdot 5 \cdot 0.1$ mm^3 plates with $\{101\}$ faces. A similar procedure carried out in Sn solution resulted in yellow-orange $15 \cdot 1 \cdot 1$ mm^3 rods of $CuGaS_2$. Although these contained about 0.2 w/o of Sn, a potential donor, no n-type conductivity was detected [5.70]. Iodine vapor transport of pre-reacted powder in closed ampoules has resulted in up to $10 \cdot 5 \cdot 1 \text{ mm}^3$ large platelets [5.39, 50–53, 71–73]. Typical conditions are 0.5 to 2 g polycrystalline material, 5 to 18 mg I_2/cm^3, 800 to 900 °C high and 600 to 750 °C low temperature, and 3 days of transport time. With exactly stoichiometric ($1.0 \text{ Cu}_2\text{S} + 1.0 \text{ Ga}_2\text{S}_3$) starting material yellow crystals were obtained, $1.1 \text{ Cu}_2\text{S} + 1.0 \text{ Ga}_2\text{S}_3$ gave black, and $1.0 \text{ Cu}_2\text{S} + 1.1$ Ga_2S_3 red crystals [5.52]. Preferred growth facets are $\{112\}$. It was noted that iodine, a potential donor, does not suppress p-type conductivity, and that it does not quench efficient green photoluminescence [5.72].

CuInS_2 has been crystallized from stoichiometric melts by cooling from 1 100 °C at a rate of 1 °C/h [5.49], or by cooling from 1 080 °C [5.74]. Solution growth from In melts was carried out as part of the investigation of CuGaS_2 and $\text{CuGa}_{1-y}\text{In}_y\text{S}_2$ growth [5.69]. CuInS_2 crystals also have been grown by conventional iodine vapor transport [5.51, 75] as well as by the vertical pulling method which resulted in $5 \cdot 5 \cdot 0.2 \text{ mm}^3$ platelets with $\{112\}$ faces [5.76]. In this method the ampoule is pulled through a V-shaped temperature profile with a maximum temperature above the nucleation point for CuInS_2.

All reported CuInS_2 crystals were prepared by directional solidification of melts [5.36, 49, 74, 77–79], or by zone leveling of polycrystalline ingots [5.32, 80]. Stoichiometric mixtures were heated to 1 050–1 100 °C and cooled at a rate of 2 to 8 °C/h. Slight excess of Se led to p-type crystals. Zone leveling of polycrystalline material prepared from the elements was carried out with cold zones either below [5.32] or above [5.80] the temperature of the tetragonal to cubic transition, 810 °C, and the hot zone at 1 050 °C.

5.2.3 Junction Formation and Preparation of Diodes

CuInS_2 and CuInSe_2 can be made n- and p-type with conductivities sufficient for preparation of homodiodes.

Homojunctions in 10 Ωcm p-CuInS_2 substrates have been obtained by two different diffusion anneals [5.76]. In one, a dot of eutectic Ga-In alloy was placed on the crystal surface. This sample was sealed in an evacuated ampoule and heated for 30 minutes at 200 °C followed by a quench to room temperature. The second method consisted in annealing p-type crystals in the presence of 0.3 mg/cm^3 InCl_3 in evacuated ampoules for 1 minutes at 600 °C. After the anneal, the ampoules were again quenched to room temperature. Both procedures led to ~ 20 μm deep junctions. It was noted that the 600 °C anneal, when carried out for longer than 5 minutes, completely converted the sample to n-type. Contact to the p-type substrate, exposed by sanding after the diffusion anneal, was made with a gold film applied by evaporation of aqueous AuCl_3 solution at ~ 60 °C. The Au film was contacted with Ag paint. Ga-In alloy served as contact to the n-type layer. Rectification ratios I(forward)/ I(reverse) at 2 V bias were $1.7 \cdot 10^4$ for diodes prepared with Ga-In anneals, and $1.5 \cdot 10^4$ when annealed in InCl_3 vapor. Zero bias resistances were $1 \cdot 10^7$ and $3 \cdot 10^7$ Ω, respectively, and forward series resistances at biases greater than 2 V amounted to 100 and 500 Ω, respectively. These comparatively large series resistances may arise from resistive layers introduced by flat diffusion profiles in the vicinity of the *pn* junctions, a consequence of fast diffusion.

Homojunctions in CuInSe$_2$ have been produced by evaporating In onto *p*-type crystals and subsequent annealing for 30 minutes at 200 °C [5.32]; by evaporating In, or Cu, onto samples held at 200 °C and maintained at this temperature for 20 minutes before cooling [5.34,38]. The indium, or copper, served as contact to the *n*-layer; low resistance ohmic contacts to the *p*-type substrates were made with silver paint applied to sandblasted areas. The diode prepared by evaporation of Cu exhibited a series resistance of 150 Ω on a cross section of 0.04 cm^2. This resistance was ascribed to the contacts. The rectification characteristic of this device was stable at 80 K, but became more ohmic after a period of days at room temperature. CuInSe$_2$ homodiodes have also been prepared from *n*-type substrates by 1 minute anneals in Se vapor at 600 to 700 °C, followed by quenching [5.81]. The anneals were carried out in sealed fused quartz ampoules at 0.1 Torr Ar pressure. The junctions were about 250 μm deep. Lower temperature anneals also produced diodes, but with excessive series resistance. Mesas etched in warm 1HCl:1HNO$_3$ from the *p*-type layer were contacted with Au, from AuCl$_3$ solution, and the *n*-substrate with a soldered In dot. At 1 V bias a rectification ratio of 300 was measured, the forward series resistance was 100 Ω at room temperature, and 5 kΩ at 77 K. 6 to 8 minute anneals at 400 °C of *p*-CuInSe$_2$ in Cd vapor within fused quartz ampoules, followed by isolation of *n*-type mesas by etching, sputtering of Au contacts to the *p*, and of In-Sn contacts to the *n*-side resulted in diode with rectification ratios of $\sim 1\,000$ [5.36]. At currents below $\sim 10^{-8}$ A, the current was linearly proportional to the forward bias; between 10^{-8} and 10^{-4} the current-voltage characteristic, $I \sim V^3$, indicated two-carrier space-charge-limited current. The value of the nearly reverse-bias independent junction capacitance corresponded to a 13 μm thick insulating layer.

Type conversion coupled to rapid diffusion in CuInSe$_2$ has been noted by several authors [5.32,36,82]. Estimated diffusion rates of the most plausible diffusing species were $D \cong 10^{-5}$ cm^{2}s^{-1} at 600 °C for Se [5.32], and $D \cong 5 \cdot 10^{-7}$ cm^{2}s^{-1} at 400 °C for Cd [5.36]. Recently, a diffusion coefficient has been deduced from measurements of the depth of *pn* junctions produced at 200 to 450 °C [5.83, Fig. 5.5]. Zn- or Cd-electroplated crystals of *p*-CuInSe$_2$ were annealed for 5 minutes in fused quartz ampoules backfilled with several torr of argon. After angle lapping the lapped faces were stained under light with 1HF:1HNO$_3$:1H$_2$O to differentiate *n*-from *p*-regions. The diffusion coefficient, $D(\text{cm}^2\text{s}^{-1}) = 164 \exp\left[-1.19(\text{eV})/kT\right]$, contains a large pre-exponential term indicating concentration or mobilities of point defects substantially above those of the related II-VI compounds.

Ion implantation followed by activation anneals has also been used to produce CuInSe$_2$ homodiodes. In one study, 10^{14} to $5 \cdot 10^{15}$ cm^{-2} Cd was implanted at 135 keV (LSS range ~ 400 Å) into *p*-CuInSe$_2$ [5.37]. These doses correspond to concentrations of $2.5 \cdot 10^{19}$ to $1 \cdot 10^{21}$ cm^{-3} in the unannealed implanted layers. 10 to 30 minute vacuum anneals were carried out at 300 to

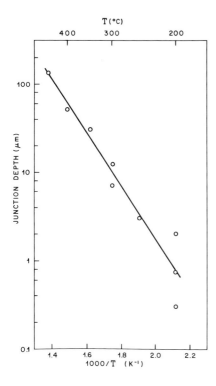

T (°C)

JUNCTION DEPTH (μm)

1000/T (K⁻¹)

Fig. 5.5. Depth of *pn*-junctions introduced in Zn- or Cd-plated *p*-CuInSe₂ by 5-minute anneals. From [5.83]

500 °C with Au contacts sputtered to the HCl:HNO₃ etched substrate, and sputtered In-Sn alloy contacts to *n*-type mesas. For a typical sample (10^{15} Cd cm^{-2}, 30 minute anneal at 370 °C), room temperature values were determined for sheet resistivity, $2.8 \cdot 10^3 \, \Omega/\square$, sheet density of free electrons, $3.9 \cdot 10^{13}$ cm^{-2}, and electron mobility, 57 cm^{2}V^{-1}s^{-1}. A sample of *p*-CuInSe₂ annealed in Cd exhibited $N_D - N_A = 2 \cdot 10^{17}$ cm^{-3}, and $\mu_e = 220$ cm^{2}V^{-1}s^{-1}. Diodes annealed at 340 °C had larger reverse breakdown voltages, ~ 30 V, and larger forward series resistances, $\sim 400 \, \Omega$, than those annealed at 400 °C (5 to 10 V and $\sim 15 \, \Omega$). The 1 MHz capacitance of 340 °C annealed diodes was nearly voltage independent; its magnitude corresponded to an insulating layer thickness of ~ 10 μm. 400 °C annealed diodes showed a rectification ratio of $\sim 2 \cdot 10^4$ at 1 V bias; the 1 MHz reverse-bias capacitance was characteristic of a linearly graded junction with $C^{-3} \sim V$. Built-in voltages lay between 0.8 and 4 V; thus some diodes did not contain a significant insulating layer. Forward current-voltage relations were typical for space charge recombination. In a second study Br, Cl, and Zn were implanted [5.35]. Implantation and annealing conditions were similar to those reported in the earlier investigation. With high doses of Zn ($1 \cdot 10^{16}$ cm^{-2}) and 350 °C anneals *p-i-n* diodes were obtained with a $I \sim V^3$ segment in the forward current-voltage characteristic, and an insulating layer of ~ 30 μm thickness. Typical room temperature values

obtained were a sheet resistivity of 140 $\Omega/\square$, a sheet density of electrons of $2 \cdot 10^{14}$ cm^{-2}, and an electron mobility of 130 cm^{2}V^{-1}s^{-1}. Br, Cl, and Zn implants annealed at 400 °C again resulted in diodes with forward current characteristic typical of recombination in the space charge. Rectification ratios at 1 V were $3 \cdot 10^5$, $2 \cdot 10^5$ and $8 \cdot 10^3$ for Br, Zn, and Cl implants, respectively. Forward series resistance was of the order of ~ 16 Ω, reverse breakdown voltages, 4 to 7 V, and the 1 MHz reverse capacitance again indicated a linearly graded junction.

Contact metals preferred in the homodiode studies described above were Au for p-type and In or Ga-In or In-Sn alloys for n-type material. Eutectic Ga-In alloys are applied at room temperature, In-Sn alloys are sputtered while pure In is soldered, or evaporated and annealed. After proper preparation of the n-type surface by etching or sanding it is usually not difficult to obtain low resistance contacts. Au-contacts to p-type compounds have posed problems, though, even when Au films are applied by electroless plating from AuCl$_3$ solution. Recent work on electroplating of Au has resulted in contacts to CuGaS$_2$, CuInS$_2$, and CuInSe$_2$ with 1 Ω or less contact resistance at room temperature on contact areas of ~ 1 mm^2. It is important that the crystal surfaces be mechanically polished and not chemically etched before plating [5.84].

Some of the most interesting junction devices based on chalcopyrites are heterodiodes. CdSnP$_2$/InP, CuGaS$_2$/CdS, and CuInSe$_2$/CdS were made to take advantage of a combination of properties not available in other semi-conductors, such as magnitude of the bandgap, conductivity type, and lattice parameter. CdSnP$_2$/InP and CuInSe$_2$/CdS are candidates for light sources and photodetectors in the wavelength regime of greatest interest for optical fiber transmission, i.e., $0.9 < \lambda < 1.4$ μm. Fibers with losses smaller than 1 dB/km over this range are now under intensive research. CuGaS$_2$/CdS is of interest for light emitting diodes in the short wavelength portion of the visible spectrum.

CdSnP$_2$ is easily prepared low-resistivity n-type but not low-resistivity p-type. Its a-parameter of 5.900 Å is close to the lattice constant of InP, 5.869 Å. The mismatch of only 0.5% suggested the feasibility of good quality hetero-junctions between these two semiconductors. CdSnP$_2$ forms peritectically from solutions of Cd and P in Sn, or of Sn and P in Cd. The habit and small size of these solution-grown crystals are inadequate for device research. Liquid-phase epitaxy of n-CdSnP$_2$ from solution onto InP substrates was an appealing alternative since large p-InP substrates can be grown. n-CdSnP$_2$/p-InP hetero-diodes were prepared from a solution of 1.5 a/o and 3.5 a/o P in Sn in a growth apparatus sealed within a fused quartz ampoule under 0.87 atm (300 K) of He [5.85–87]. The melt was contained in a vitreous carbon crucible. The solution of Cd and P in Sn was prehomogenized for 1 hour at 600 °C. When solidified, it consists of a dispersion of small CdSnP$_2$ crystals in Sn. It is loaded into the growth apparatus, homogenized for 15 minutes at 525 °C, and rapidly cooled to 510 °C before tipping. After tipping, the furnace is cooled at a rate of 10 °C/h. The melt was made to flow slowly over the (100) InP substrate to feed the growth area continuously with fresh solution. Typically, 0.15 mm thick $CdSnP_2$ layers were grown on 0.4 mm thick substrates. Sn adhering to the finished heterostructure was removed by Hg extraction followed by etching in HF/HNO_3, and by a Br_2/methanol polish. The initial $1 \cdot 1.5$ cm^2 wafers were cleaved along the InP (110) planes to $1 \cdot 1$ mm^2 diodes. Ohmic contacts to the $1.2 \cdot 10^{16}$ to $5 \cdot 10^{18}$ Zn or Cd doped InP substrates were made with spot-welded 5% Zn-doped Au wires, and In was soldered to $CdSnP_2$. The LPE-grown $CdSnP_2$ had $N_D - N_A$ of $\sim 5 \cdot 10^{18}$ cm^{-3} and $\mu_e \sim 20$ cm^{2}V^{-1}s^{-1}. The electron mobility in nominally pure solution-grown $CdSnP_2$ with $N_D - N_A$ $\sim 1 \cdot 10^{17}$ cm^{-3} is typically 2 000 cm^{2}V^{-1}s^{-1}. InP precipitates, inferred from electron and ion microprobe analysis, are responsible for the lower mobility in the epitaxial layers. Diodes prepared on a $5 \cdot 10^{17}$ cm^{-3} Cd-doped InP substrate had a rectification ratio of 100 to 0.7 V bias, a 0.2 MΩ zero-bias resistance, and a linear $C^{-2} - V$ characteristic typical of a one-sided junction, with a slope corresponding to $4 \cdot 10^{17}$ cm^{-3} acceptor density. Diodes on a $1.2 \cdot 10^{16}$ cm^{-3} Zn-doped InP substrate were more rectifying with a ratio of 10^5 at 1 V bias, and a zero-bias resistance of 100 MΩ. They again exhibited the $C - V$ characteristics of a one-sided junction with the acceptor density of the initial substrate.

p-$CuGaS_2$/n-CdS heterodiodes have been prepared by growth of CdS films on $CuGaS_2$ substrates [5.51, 88–90], and by deposition of $CuGaS_2$ films on CdS substrates [5.90]. These devices were used for an investigation of the feasibility of light-emitting diodes constructed from the exclusively n-type large gap II-VI semiconductors, and the exclusively p-type large gap I-III-VI$_2$ chalcopyrites. The rather large lattice mismatch of $\sim 10\%$ between $CuGaS_2$ and CdS which precludes efficient carrier injection was considered less important than the high conductivity attainable in CdS films. n-CdS films were grown in vacuum stations either from the elemental vapors [5.51, 88, 89] or by flash evaporation of CdS powder [5.90]. Melt-grown $CuGaS_2$ substrates of the green variety with natural (112) faces were polished with 3 μm diamond grit, annealed in excess sulfur for 24 hours at 800 °C and quenched to obtain crystals with $\rho \cong 1$ Ω cm, $\mu_e = 15$ cm^{2}V^{-1}s^{-1}, $N_A - N_D = 5 \cdot 10^{17}$ cm^{-3}. Some vapor-transported $CuGaS_2$ substrates were used in the flash-evaporation study [5.90]. After the sulfur anneal, the substrates were cleaned in carbon disulfide, etched in 50 °C 1HCl:1HNO$_3$, rinsed in distilled water and finally in methanol. Since substrate temperatures of higher than 130 °C were observed to produce resistive layer at the surface of $CuGaS_2$, growth temperatures were kept between 100 and 130 °C. Molecular beam grown CdS films were about 5 μm thick with $\rho \cong 1$ Ω cm, and grew epitaxially on the $CuGaS_2$ substrate [5.89]. The resistivity of flash-evaporated CdS with thicknesses of 0.5 to 1 μm ranged from 1 to 10 Ωcm [5.90]. Electroless [5.51, 88, 89] or evaporated [5.90] Au contacts were applied to $CuGaS_2$, and In, Ga-In, Hg-In, or Ga-In-Sn contacts to CdS. Current-voltage characteristics were typical of one-carrier space-charge limited flow. In a typical sample [5.89] the forward current was linearly proportional to the sixth power of the forward voltage. These characteristics were ascribed to a resistive layer with a thickness of 5 μm or more, on the CuGaS$_2$ side of the interface. p-CuGaS$_2$/n-CdS and p-CuGaS$_2$/n-ZnSe hetero-structures have also been prepared by flash evaporation of CuGaS$_2$ onto the respective binary substrates [5.90]. Rectifying diodes were obtained in both cases; the p-CuGaS$_2$/n-CdS structure proved a true heterodiode, while the $I-V$ characteristic of the ZnSe based devices was dominated by a metal-insulator-semiconductor diode composed of ZnSe, insulating ZnSe, and Au contacts. Diodes with evaporated CuGaS$_2$ films did not produce electro-luminescence.

The p-CuInSe$_2$/n-CdS heterodiode [5.82,91,92] takes advantage of the small lattice mismatch ($\sim 1.2\%$) between the two partners. 5 to 10 μm thick n-CdS films were grown from the elemental vapors in the simple molecular beam apparatus used for CuGaS$_2$/CdS diodes. Melt-grown CuInSe$_2$ substrates were annealed for 24 hours at 600 °C in selenium vapor to obtain p-type con-ductivity with $\rho \cong 0.5$ Ωcm. Substrates were kept at 130 °C during film growth to prevent excessive relaxation of the acceptor density in the surface of CuInSe$_2$. Electroless Au contacts were applied to CuInSe$_2$, and Ga-In or In contacts to CdS. When diodes are selected such that they are free of microcracks in the CuInSe$_2$ substrate, the reverse resistance at 1 V bias reaches ~ 10 MΩ for 1 mm^2 area. $C-V$ characteristics indicate an abrupt heterojunction with an acceptor density in the CuInSe$_2$ considerably below that of the bulk substrate.

On some CuInSe$_2$ crystals it was possible to differentiate between the (112), A, or metal, face and the ($\bar{1}\bar{1}2$), B, or selenium face by etching in a 0.5 to 5 vol. % Br$_2$-methanol solution [5.82]. One face etches shiny, the other mat. By analogy to etching of ZnGeP$_2$ and of AgGaS$_2$ [5.93], the shiny face was tentatively identified as the A face.

All-thin-film p-CuInSe$_2$/n-CdS heterodiodes have been grown, during a program on photovoltaic converters, by evaporation of CuInSe$_2$ onto CdS films, and be evaporation of CdS onto CuInSe$_2$ films [5.94].

Several Schottky barrier and MIS diodes based on chalcopyrites have been reported. Among these are a CuInSe$_2$ MIS diode used for low-temperature electroreflectance measurements [5.18] and a p-CuInSe$_2$/Au photodetector [5.38]. An electroluminescent p-CuGaS$_2$/In diode has been prepared by evaporation of ohmic Au contacts and of rectifying In contacts onto 10^2–10^3 Ωcm green CuGaS$_2$ grown by iodine transport [5.95].

5.3 Luminescent Transitions and Electroluminescence

5.3.1 Photoluminescence and Cathodoluminescence

Both spontaneous and stimulated photo- and cathodoluminescence have been observed in CdSnP$_2$. When nominally undoped crystals (0.01 to 0.1 Ωcm n-type) are excited with a 0.6328 μm He-Ne laser at 1.7 K ($E_g = 1.240$ eV), the luminescence is dominated by a band consisting of three narrow lines at

1.0032, 1.0038, and 1.0046 µm. These lines, which merge at an excitation density higher than ∼ 50 mW/cm², are ascribed to the recombination of bound excitons. Weaker, and broader, bands at 1.03, 1.05, and 1.075 µm, the latter possibly made up of two overlapping bands, are thought to arise from recombination through impurity levels [5.96,97]. The bound exciton lines are observed only at selected spots of the crystals. Ag-doped crystals photoluminesce uniformly in four narrow bands at 1.004, 1.026, 1.072, and 1.124 µm (1.7 K). With increasing excitation by an argon laser the 1.004 µm line grows linearly while the luminescence at longer wavelengths saturates. At yet higher pumping levels ($\sim 10^5$ W/cm²) available with a pulsed nitrogen laser (0.3371 µm) a stimulated emission line appears with a peak at 1.008 to 1.015 µm (Fig. 5.6a). A laser (Fig. 5.6b) with a threshold of ∼ 50 kW/cm² can be fabricated using the natural (112) faces to form the Fabry-Perot cavity, and pumping through a cleaved ($1\bar{1}0$) face [5.98]. The spontaneous emission at 1.004 µm is ascribed to band-to-band recombination. From studies of the luminescence spectra in a magnetic field it was concluded that the stimulated emission, although 5 to 10 meV lower in energy, arises from the same mechanism. The other spontaneous emission lines of Ag-doped CdSnP₂ are though to result from recombination through levels involving Ag. Photoluminescence of Cu-

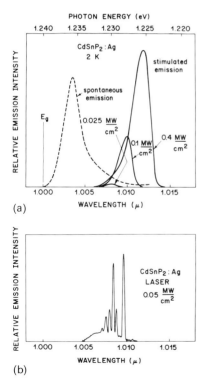

(a)

(b)

Fig. 5.6. (a) Spontaneous and stimulated emission spectra of Ag-doped CnSnP₂ excited with a pulsed nitrogen laser. (b) Laser emission spectrum of Ag-doped CdSnP₂ pumped with a pulsed nitrogen laser slightly above threshold. From [5.98] doped $CdSnP_2$ is characterized by a band ~ 130 meV below the gap, possibly from a conduction band to acceptor transition, and three or four weaker bands closer to the gap energy. It is interesting to note that Cu-doped $CdSnP_2$ (~ 10 Ωcm p-type) and undoped $CdSnP_2$ (0.01 to 0.1 Ωcm n-type) are comparatively inefficient in luminescence, while Ag-doped $CdSnP_2$ (1 Ωcm n-type) with $\sim 10^{18}$ cm^{-3} Ag but $N_D - N_A \cong 10^{17}$ cm^{-3} is an efficient medium [5.98]. One explanation could be that the addition of Ag to undoped $CdSnP_2$ reduces the density of sites promoting non-radiative recombination.

Spontaneous luminescence from electron beam pumped undoped $CdSnP_2$ at 80 K peaks at 1.01 μm with tails extending to 1.03 and to 0.99 μm [5.99]. An investigation at 80 and 300 K revealed that at low excitation levels (0.2 Acm^{-2}) the spectra are dominated by low energy emission (~ 1.14 eV at 80 K, ~ 0.92 eV at 300 K). At electron currents near 1 Acm^{-2} the principal spectral feature is a peak close to bandgap energy [5.100]. In experiments with a pulsed electron beam at 1.7 K, the lifetime for the 1.004 μm band-to-band emission was estimated shorter than 5 ns, while the broader bands ascribed to impurity levels decayed with lifetimes from 100 to 1000 ns [5.96]. 25 keV electron-beam pumped Fabry-Perot resonators with polished or cleaved faces had a lasing threshold of 0.2 Acm^{-2} and radiated at a peak of 1.0139 μm. The current threshold was estimated to be equivalent to a current density of 1200 Acm^{-2} in a pn junction injection laser [5.99].

Aside from a brief cathodoluminescence study [5.101], $CuGaS_2$ has been investigated exclusively by photoluminescence [5.102, 103]. The 2 K spectrum is characterized by recombination of free and bound excitons with $\lambda = 0.4953$ μm and 0.4968 μm. In melt-grown crystals intense edge emission is observed near 0.520 μm. It is probably associated with donor-acceptor recombination [5.103]. $CuGaS_2$ crystals grown by iodine transport, and thus doped with iodine, exhibit efficient green photoluminescence (2 K, 0.4416 μm HeCd laser) with the most intense and highest energy line at 0.5163 μm, followed by a large number of equidistant lines separated by 4.6 meV, possibly the local mode sidebands of a bound exciton recombination. Anneals in sulfur lead to a broad band with $0.505 < \lambda < 0.550$ μm [5.72]. In-diffusion of Zn or Cd shifts the photoluminescent peak and efficiency to higher wavelengths [5.51] with the internal quantum efficiency reaching 0.2 percent. Stimulated emission at 2 K has been excited in selected green melt-grown crystals with a pulsed nitrogen laser (0.3371 μm). For pump intensities above ~ 1 MWcm^{-2} a superlinearly increasing line near 0.500 μm appears. The emitted photons have an energy slightly less than a free exciton. No resonant modes were observed [5.102]. The photoluminescence of iron-doped vapor-transported samples has also been studied [5.104].

The photoluminescence spectrum of $CuInS_2$ at 2 K exhibits a structure very similar to that of $CuGaS_2$. A peak at 0.8075 μm originates from free exciton recombination, several bound exciton emissions follow at lower energy. Broad bands at 0.86 to 0.89 μm probably result from donor-acceptor recombination [5.103]. In a cathodoluminescence study of $CuInSe_{2-y}S_y$ alloys

$(0 < y < 2)$ at 90 to 300 K a similar broad band was observed in all alloys, including the end members. It is attributed to recombination via donor-acceptor pairs or via distributed levels with a low density of states [5.74]. Pumping with a pulsed nitrogen laser as described above for $CuGaS_2$ led to stimulated emission in $CuInS_2$ at 0.8175 µm (2 K), probably involving impurities or defects [5.102]. The electron-beam pumped stimulated emission at 80 K appeared in the long-wavelength wing of a narrow 1.519 eV band that is observed at high exitation density only. Like the laser-induced stimulated emission, the electron-beam stimulated emission is due to an impurity-band transition. The threshold for single-pass amplification (no resonator) was ~ 15 A cm^{-2} [5.74].

The first photoluminescence spectra taken of $CuInSe_2$ consisted of a single band at ~ 0.99 eV at 2 K, and of a weak broad band between 0.95 and 1.05 eV at 77 K [5.105]. Since the observation of electroluminescence in this compound several studies have been concerned with the nature of radiative recombination in $CuInSe_2$. The low temperature spectra of as-grown or sulfur- or vacuum-annealed crystals are characteristic of the conductivity type. At 77 K, p-$CuInSe_2$ luminescences in a band at 1.00 eV and n-$CuInSe_2$ at 0.93 eV. These spectra can be interchanged by alternate anneals in minimum Se pressure (to introduce n-type conductivity) or in maximum Se pressure (to introduce p-type conductivity). In some high-resistivity samples a band-to-band emission at 1.04 eV is detected. The 1.00 eV luminescence is attributed to free-to-bound transition to an acceptor level ($\varepsilon_A \cong 40$ meV), and the 0.93 eV radiation to a donor-acceptor pair recombination with $\varepsilon_D \cong 70$ meV. This latter luminescence was also shown by Cd-diffused $CuInSe_2$ while Zn-diffused samples exhibited a very broad band with a peak shifted to lower energies (0.85 eV at 77 K) [5.33]. In two papers on luminescence from p-type and Cd-implanted n-type $CuInSe_2$, a broad band peaking at 0.94 to 0.95 eV (2 K) was attributed to donor-acceptor pair recombination. Copper vacancies are thought to be the donors with $\varepsilon_D = 65 \pm 2$ meV, and selenium vacancies the acceptors, $\varepsilon_A = 85 \pm 2$ meV. Cd-implanted samples are characterized by a broad band peaking at ~ 0.89 eV (2 K) [5.79,106]. The cathodoluminescence spectrum of $CuInSe_2$ was found similar to that of $CuInS_2$, and stimulated emission was obtained at 0.993 eV (90 K) [5.74].

5.3.2 Electroluminescence

$CuInS_2$ homodiodes, prepared by annealing p-$CuInS_2$ with a Ga-In dot, luminesce in two bands peaking at 1.40 and more intensely at 1.48 eV at 300 K ($E_g = 1.53$ eV), and in one band at 1.42 eV at 77 K ($E_g = 1.55$ eV) (Fig. 5.7). This 1.42 eV peak also dominates the 77 K photoluminescence spectrum. The internal quantum efficiency, calculated using a refractive index of $n = 2.8$ for $CuInS_2$, is 10^{-5} at 300 K and 10^{-3} at 77 K [5.76].

$CuInSe_2$ homodiodes prepared by introducing either a p-layer in a n-substrate or a n-layer in a p-substrate show identical spectra at 77 K ($E_g = 1.04$ eV).

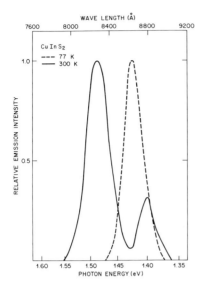

Fig. 5.7. Electroluminescence spectra of a CuInS$_2$ homodiode at 77 and 300 K. From [5.76]

They consist of one band peaking at 0.92 to 0.93 eV. The internal quantum efficiency is $\sim 10^{-3}$ at 300 K and $\sim 10^{-1}$ at 77 K. The electroluminescence is attributed to donor ($\varepsilon_D \cong 70$ meV)—acceptor ($\varepsilon_A \cong 40$ meV)—pair recombination [5.33,81].

In diodes prepared by Cd-ion implantation of p-CuInSe$_2$ followed by an activation anneal single bands appeared with peak energies ranging from 0.92 to 0.98 eV (77 K). The peak position was either independent of, or increased with diode current. Best internal quantum efficiencies were $\sim 10^{-3}$ at room temperature and ~ 0.15 at 77 K. At this temperature the intensity of the luminescence was proportional to the square of the forward current indicating recombination in the depletion layer of the diode [5.37].

Br-, Cl-, and Zn-implanted and annealed CuInSe$_2$ homodiodes also luminesce with peaks ranging from 0.92 to 0.95 eV (77 K). These bands are ascribed to donor-acceptor-pair recombination with a Cu vacancy acceptor and the implanted donor participating. The spread in peak energies of different samples implanted with the same donor element is explained with different Fermi levels due to varying degrees of compensation. The Cl-implanted diode shows a second peak at 1.01 eV which is ascribed to a Cl donor-to-valence band transition. Highest external quantum efficiency obtained at 77 K is $8 \cdot 10^{-3}$, and best internal quantum efficiencies are $\sim 10^{-3}$ at room temperature and ~ 0.2 at 77 K. The luminescent intensity is again proportional to the square of the forward current [5.35].

n-CdSnP$_2$/p-InP heterodiodes prepared by LPE growth of CnSnP$_2$ on InP substrates electroluminesce in a very broad-band peaking at ~ 0.8 eV (300 K) and at ~ 0.9 eV at low temperatures. Since the 300 K band gaps of CdSnP$_2$ and InP are 1.17 and 1.34 eV, respectively, the emission probably originates in an alloy with reduced band gap at the $CdSnP_2/InP$ interface. Room temperature internal quantum efficiencies are typically $1 \cdot 10^{-2}$ although they can reach $2 \cdot 10^{-2}$. External quantum efficiencies are ~ 25 times smaller. The efficiency is approximately constant at temperatures up to ~ 150 K and then drops exponentially with $1/T$ with a slope corresponding to 68 meV [5.85–87] (Fig. 5.8).

p-$CuGaS_2/n$-CdS heterodiodes prepared by coevaporation of Cd and S onto $CuGaS_2$ substrates electroluminesce with a peak at 2.30 eV (300 K) and at 2.40 (77 K), respectively. Comparison of the electroluminescence spectrum with the photoluminescence spectrum of the $CuGaS_2$ substrate and of the CdS layer demonstrates that the electroluminescence originates in the $CuGaS_2$ (Fig. 5.9); it is a result of electron injection into the $CuGaS_2$. The green light intensity rises very steeply with direct forward current and, at higher current density, linearly with pulsed forward current. External quantum efficiencies are 10^{-5} at room temperature and 10^{-3} at 77 K. Between 77 K

Table 5.2. Main peak emission energies E_p and quantum efficiencies QE of chalcopyrite-based light emitting diodes

	E_p (eV)		E_g (eV)		Internal QE		External QE		Ref.
	77 K	300 K	77 K	300 K	77 K	300 K	77 K	300 K	
pn CuInS$_2$	1.42	1.48	1.55	1.53	10^{-3}	10^{-5}			[5.76]
pn CuInSe$_2$	0.92–0.98		1.04	1.01	0.2	10^{-3}	$8 \cdot 10^{-3}$		[5.33, 35, 37, 81]
n CdSnP$_2$/pInP	~ 0.9	~ 0.8		1.17	$8 \cdot 10^{-2}$	$2 \cdot 10^{-2}$	$3 \cdot 10^{-3}$	$8 \cdot 10^{-4}$	[5.85–87]
p CuGaS$_2$/nCdS	2.40	2.30	2.50	2.40			10^{-3}	10^{-5}	[5.88, 89]
p CuInSe$_2$/nCdS		0.89		1.01			10^{-2}	10^{-4}	[5.82]

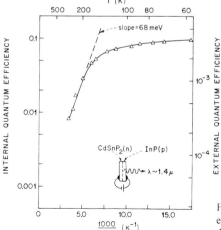

Fig. 5.8. Temperature dependence of the quantum efficiency of a $CdSnP_2/InP$ light-emitting heterodiode. From [5.85]

and room temperature the peak energy follows the bandgap energy with $E_{gap} - E_{peak} = 100$ meV. It is speculated that donor ($\varepsilon_D \cong 30$ meV)—acceptor ($\varepsilon_A \cong 70$ meV) pair recombination is the light-producing mechanism. A likely donor is Cd which has diffused into the CuGaS$_2$ during CdS growth. The green band is usually composed of three peaks separated by 45 meV, suggesting phonon replication of the zero phonon line. The 2.40 eV peak dominates at high excitation density while the 2.31 eV peak is prominent at low excitation [5.88,89].

When CuGaS$_2$/CdS heterodiodes are prepared by flash evaporation of one component on substrates of the other, only diodes with evaporated CdS on bulk CuGaS$_2$ are found to luminesce. This supports the conclusion that electroluminescence originates in the CuGaS$_2$ which has to be of sufficient crystalline perfection to permit radiative recombination. The luminescent peak in these diodes lies at 2.42 eV (77 K) with side bands at 2.39 and 2.34 eV and an intense broad-band between ~ 2.25 and ~ 1.6 eV [5.90].

Deep electroluminescence was observed from a In/p-CuGaS$_2$ Schottky barrier diode. At liquid nitrogen temperature this spectrum is characterized by three broad-bands at ~ 1.85, ~ 1.45, and ~ 1.25 eV. The intensity of these bands increases with decreasing peak energy. A very weak peak near the bandgap energy of CuGaS$_2$ is also observed. The emission intensity is linearly proportional to $I(\text{forward})^{1.4}$ [5.95].

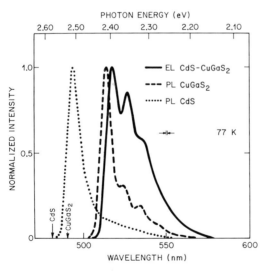

Fig. 5.9. Photoluminescence (PL) spectra of the CdS layer and of a bare CuGaS$_2$ substrate, and the DC electroluminescence (EL) spectrum of a CuGaS$_2$/CdS diode, all at 77 K. Bandgaps are marked by arrows. From [5.88]

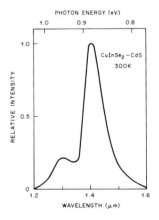

Fig. 5.10. Electroluminescence spectrum of a CuInSe$_2$/CdS heterodiode. From [5.82]

p-CuInSe$_2$/n-CdS heterodiodes prepared by coevaporation of Cd and S onto CuInSe$_2$ substrates luminesce at room temperature with a peak energy of ~ 0.89 eV and a low intensity band at 0.96 eV (Fig. 5.10). These peaks possibly involve donor-acceptor pair, and conduction band-to-acceptor recombination, respectively. External quantum efficiencies are $\sim 10^{-4}$ at room temperature and $\sim 10^{-2}$ at 77 K [5.82,91].

5.4 Conclusion

The expectation that chalcopyrites support electroluminescence at near-band-gap energies has been borne out by the diodes prepared to date. Experience accumulated with these devices also points to the need for more detailed knowledge of, and more precise control over, material and electronic properties. Further work on electroluminescence and other applications, as well as fundamental studies, will lead to the comprehensive understanding indispensable for commercial exploitation. Of course, a more concerted development effort can be expected once a chalcopyrite is found to perform better than its binary counterpart. Development of such alternatives is the principal task of today's chalcopyrite device research.

The author would like to acknowledge the help rendered by Dr. *B. Tell* and Dr. *J. L. Shay* through numerous discussions.

References

5.1   N. A. Goryunova: *The Chemistry of Diamond-Like Semiconductors* (MIT Press, Cambridge, Mass. 1965)
5.2   N. A. Goryunova: *Slozhnye Almazopodobnye Poluprovodniki* (Sovietskoye Radio, Moscow 1968)
5.3   L. I. Berger, V. D. Prochukhan: *Ternary Diamond-Like Semiconductors* (Consultants Bureau, New York 1969)
5.4   R. C. J. Draper, A. Miller, R. G. Humphreys: *A Bibliography and Guide to Literature of II-IV-V$_2$ Ternary Compounds* (School of Physics, University of Bath, England 1973); to be updated in 1977
5.5   *Poluprovodniki* A^{2}B^{4}C$^5_2$, ed. by N. A. Goryunova, Yu. A. Valov (Sovietskoye Radio, Moscow 1974)
5.6   U. Kaufmann, J. Schneider: In *Festkörperprobleme – Advances in Solid State Physics,* Vol. 14 (Vieweg-Pergamon, Braunschweig 1974) p. 229
5.7   R. C. J. Draper, C. Schwab: *Bibliographie et Guide de la Littérature des Composés* I-III-VI$_2$ (University of Bath, England, and Laboratoire de Spectroscopie et d'Optique du Corps Solide, Université Louis Pasteur, Strasbourg, France 1975)
5.8   "Seconde Conférence Internationale sur les Composés Semiconducteurs Ternaires, Strasbourg 1975". J. Phys. (Paris) **36**, Supplément C-3 (1975)
5.9   J. L. Shay, J. H. Wernick: *Ternary Chalcopyrite Semiconductors – Growth, Electronic Properties, and Applications* (Pergamon Press, New York 1975)
5.10 M. Kokta, J. R. Carruthers, M. Grasso, H. M. Kasper, B. Tell: J. Electron. Mater. **5**, 69 (1976)
5.11 J. Schneider, P. W. Yu: Aerospace Research Laboratories Report ARL TR 74–0004, January 1974, NTIS Access No. AD 778413

5.12   S. C. Abrahams, J. L. Bernstein: J. Chem. Phys. **55**, 796 (1971)
5.13   Y. Montfort, J. Vizot, A. Deschanvres: phys. stat. sol. (a) **29**, 551 (1975)
5.14   G. F. Karavaev, A. S. Poplavnoi, V. A. Chaldyshev: Sov. Phys. – Semicond. **2**, 93 (1968)
5.15   J. L. Shay, B. Tell, E. Buehler, J. H. Wernick: Phys. Rev. Lett. **30**, 983 (1973)
5.16   J. N. Gan, J. Tauc, V. G. Lambrecht, Jr., M. Robbins: Solid State Commun. **15**, 605 (1974)
5.17   J. E. Rowe, J. L. Shay: Phys. Rev. B**3**, 451 (1971)
5.18   J. L. Shay, H. M. Kasper: Phys. Rev. Lett. **29**, 1162 (1972)
5.19   M. J. Luciano, C. J. Vesely: Appl. Phys. Lett. **23**, 60, 453 (1973)
5.20   S. Kono, M. Okusawa: J. Phys. Soc. Japan **37**, 1301 (1974)
5.21   W. Braun, A. Goldmann, M. Cardona: Phys. Rev. B**10**, 5069 (1974)
5.22   K. D. Becker: unpublished
5.23   B. Tell, P. M. Bridenbaugh: Phys. Rev. B**12**, 3330 (1975)
5.24   A. S. Poplavnoi, Yu. I. Polygalov: Izv. Akad. Nauk SSSR, Neorg. Mat. **7**, 1706 (1971)
5.25   B. Tell, J. L. Shay, H. M. Kasper: J. Appl. Phys. **43**, 2469 (1972)
5.26   H. J. von Bardeleben, A. Goltzéné, C. Schwab: phys. stat. sol. (b) **76**, 363 (1976)
5.27   E. I. Leonov, V. M. Orlov, V. I. Sokolova, Yu. G. Shreter: phys. stat. sol. (a) **8**, 387 (1971)
5.28   E. Buehler, J. H. Wernick, J. L. Shay: Mat. Res. Bull. **6**, 303 (1971)
5.29   B. Tell, H. M. Kasper: J. Appl. Phys. **44**, 4988 (1973)
5.30   P. W. Yu, D. L. Downing, Y. S. Park: J. Appl. Phys. **45**, 5283 (1974)
5.31   D. C. Look, J. C. Manthuruthil: J. Phys. Chem. Sol. **37**, 173 (1976)
5.32   J. Parkes, R. D. Tomlinson, M. J. Hampshire: J. Cryst. Growth **20**, 315 (1973)
5.33   P. Migliorato, J. L. Shay, H. M. Kasper, S. Wagner: J. Appl. Phys. **46**, 1777 (1975)
5.34   R. D. Tomlinson, E. Elliott, J. Parkes, M. J. Hampshire: Appl. Phys. Lett. **26**, 383 (1975)
5.35   P. W. Yu, Y. S. Park, J. T. Grant: Appl. Phys. Lett. **28**, 214 (1976)
5.36   P. W. Yu, S. P. Faile, Y. S. Park: Appl. Phys. Lett. **26**, 384 (1975)
5.37   P. W. Yu, Y. S. Park, S. P. Faile, J. E. Ehret: Appl. Phys. Lett. **26**, 717 (1975)
5.38   J. Parkes, R. D. Tomlinson, M. J. Hampshire: Solid-State Electron. **16**, 773 (1973)
5.39   J. Schneider, A. Räuber, G. Brandt: J. Phys. Chem. Sol. **34**, 443 (1973)
5.40   H. J. von Bardeleben, A. Goltzéné, B. Meyer, C. Schwab: J. Phys. (Paris) Suppl. **35**, 165 (1974)
5.41   H. J. von Bardeleben, A. Goltzéné, C. Schwab: J. Phys. (Paris) **36**, C 3–47 (1975)
5.42   U. Kaufmann: Phys. Rev. B**11**, 2478 (1975)
5.43   H. J. von Bardeleben, C. Schwab, A. Goltzéné: Phys. Lett. **51** A, 460 (1975)
5.44   G. L. Troeger, R. N. Rogers, H. M. Kasper: J. Phys. C**9**, L 73 (1976)
5.45   H. J. von Bardeleben, A. Goltzéné, C. Schwab, J. M. Friedt, R. Poinsot: J. Appl. Phys. **46**, 1736 (1975)
5.46   V. I. Sokolova: *Synthesis and Investigation of Complex Semiconducting Phosphides*, Ph. D. Thesis (I. V. Grebenshchikov, Institute of Silicate Chemistry of the USSR Academy of Science, 1969) [Ref. 5.5, p. 193]
5.47   H. Hahn, G. Frank, W. Klingler, A. Meyer, G. Störger: Z. Anorg. Allg. Chem. **271**, 153 (1953)
5.48   E. K. Belova, V. M. Koshkin, L. S. Palatnik: Izv. Akad. Nauk SSSR, Neorg. Mat. **3**, 617 (1967)
5.49   H. M. Kasper: In *Reactivity of Solids*, Proc. 7th Intern. Symp. Reactivity of Solids, Bristol, 17–21 July 1972 (Chapman and Hall, London 1972) p. 46
5.50   J. L. Regolini, S. Lewonczuk, J. Ringeissen, S. Nikitine, C. Schwab: phys. stat. sol. (b) **55**, 193 (1973)
5.51   J. L. Shay, P. M. Bridenbaugh, H. M. Kasper: J. Appl. Phys. **45**, 4491 (1974)
5.52   N. Yamamoto, N. Tohge, T. Miyauchi: Japan. J. Appl. Phys. **14**, 192 (1975)
5.53   W. R. Cook, Jr.: *Growth of Large Band Gap* I-III-VI$_2$ *Crystals* (Gould, Inc., Cleveland, Ohio 1973); NTIS Access No. AD-773180
5.54   T. Teranishi, K. Sato, K. Kondo: J. Phys. Soc. Japan **36**, 1618 (1974)
5.55   K. Sato, T. Teranishi: J. Phys. Soc. Japan **37**, 415 (1974)
5.56   H. M. Kasper: Proc. 5th Materials Research Symp., NBS Special Publ. No. 364, 1972, p. 671
5.57   L. S. Palatnik, E. I. Rogacheva: Sov. Phys. – Doklady **12**, 503 (1967)

5.58 M. Robbins, M. A. Miksovsky: J. Sol. State Chem. **5**, 462 (1972)

5.59 M. Robbins, V. G. Lambrecht, Jr.: J. Sol. State Chem. **6**, 402 (1973)

5.60 M. Robbins, V. G. Lambrecht, Jr.: Mat. Res. Bull. **8**, 703 (1973)

5.61 M. Robbins, J. C. Phillips, V. G. Lambrecht, Jr.: J. Phys. Chem. Sol. **34**, 1205 (1973)

5.62 M. DiGiuseppe, J. Steger, A. Wold, E. Kostiner: Inorg. Chem. **13**, 1828 (1974)

5.63 G. V. Loshakova, R. L. Plyechko, A. A. Vaipolin, B. V. Pavlov, Yu. A. Valov, N. A. Goryuno-va: Izv. Akad. Nauk SSSR, Neorg. Mater. **2**, 1966 (1966)

5.64 A. J. Springthorpe, B. R. Pamplin: J. Cryst. Growth **3**, 313 (1968)

5.65 V. I. Sokolova, V. M. Orlov, G. P. Shpen'kov, et al.: "Some Optical Properties of CdSnP$_2$ Single Crystals". In *Physics Abstracts of the Papers at the 26th Scientific Conference*, LISI, 1968, p. 14 [Ref. 5.5, p. 118]

5.66 G. V. Loshakova: *Preparation of Crystals of the Semiconducting Compound* ZnSnP$_2$ *and Investigation of Their Properties*, Ph. D. Thesis (A. I. Gertsen, Pedagogical Institute, Lenin-grad 1969) p. 159 [Ref. 5.5, p. 118]

5.67 E. Buehler, J. H. Wernick: J. Cryst. Growth **8**, 324 (1971)

5.68 E. F. Apple: J. Electrochem. Soc. **105**, 251 (1958)

5.69 N. Yamamoto, T. Miyauchi: Japan. J. Appl. Phys. **11**, 1383 (1972)

5.70 R. H. Plovnick: Mat. Res. Bull. **10**, 555 (1975)

5.71 W. N. Honeyman, K. H. Wilkinson: J. Phys. D **4**, 1182 (1971)

5.72 J. L. Shay, P. M. Bridenbaugh, B. Tell, H. M. Kasper: J. Luminesc. **6**, 140 (1973)

5.73 H. W. Spiess, U. Haeberlen, G. Brandt, A. Räuber, J. Schneider: phys. stat. sol. (b) **62**, 183 (1974)

5.74 A. I. Dirochka, G. S. Ivanova, L. N. Kurbatov, E. V. Sinitsyn, F. F. Kharakhorin, E. N. Kho-lina: Sov. Phys.-Semicond. **9**, 742 (1975)

5.75 G. Brandt, A. Räuber, J. Schneider: Sol. State Commun. **12**, 481 (1973)

5.76 P. M. Bridenbaugh, P. Migliorato: Appl. Phys. Lett. **26**, 459 (1975)

5.77 I. G. Austin, C. H. L. Goodman, A. E. Pengelly: J. Electrochem. Soc. **103**, 609 (1956)

5.78 V. P. Zhuze, V. M. Sergeeva, E. L. Shtrum: Sov. Phys.-Techn. Phys. **3**, 1925 (1958)

5.79 P. W. Yu: J. Appl. Phys. **47**, 677 (1976)

5.80 P. M. Bridenbaugh: unpublished

5.81 P. Migliorato, B. Tell, J. L. Shay, H. M. Kasper: Appl. Phys. Lett. **24**, 227 (1974)

5.82 S. Wagner, J. L. Shay, H. M. Kasper: J. Phys. (Paris) **36**, C3–101 (1975)

5.83 B. Tell, S. Wagner, P. M. Bridenbaugh: Appl. Phys. Lett. **28**, 454 (1976)

5.84 B. Tell, P. M. Bridenbaugh: J. Appl. Phys., to be published

5.85 J. L. Shay, K. J. Bachmann, E. Buehler, J. H. Wernick: Appl. Phys. Lett. **23**, 226 (1973)

5.86 J. L. Shay, K. J. Bachmann, E. Buehler: J. Appl. Phys. **45**, 1302 (1974)

5.87 K. J. Bachmann, E. Buehler, J. L. Shay, G. W. Kammlott: J. Electron. Mater. **3**, 451 (1974)

5.88 S. Wagner, J. L. Shay, B. Tell, H. M. Kasper: Appl. Phys. Lett. **22**, 351 (1973)

5.89 S. Wagner: J. Appl. Phys. **45**, 246 (1974)

5.90 D. L. Boesen: *Optical and Electrical Properties of* CuGaS$_2$/CdS *Junctions*, Thesis (Air Force Institute of Technology, Wright-Patterson Air Force Base, Ohio, February 1974); NTIS Access No. AD-777840

5.91 S. Wagner, J. L. Shay, P. Migliorato, H. M. Kasper: Appl. Phys. Lett. **25**, 434 (1974)

5.92 J. L. Shay, S. Wagner, H. M. Kasper: Appl. Phys. Lett. **27**, 89 (1975)

5.93 S. C. Abrahams, R. L. Barns, J. L. Bernstein, E. H. Turner: Sol. State Commun. **15**, 737 (1974)

5.94 L. L. Kazmerski, F. R. White, G. K. Morgan: Appl. Phys. Lett. **29**, 268 (1976)

5.95 C. Paorici, N. Romeo, G. Sberveglieri, R. Braglia: J. Luminesc. **9**, 71 (1974)

5.96 J. L. Shay, R. F. Leheny, E. Buehler, J. Wernick: Appl. Phys. Lett. **16**, 357 (1970)

5.97 J. L. Shay, W. D. Johnston, Jr., E. Buehler, J. H. Wernick: Phys. Rev. Lett. **27**, 711 (1971)

5.98 J. L. Shay, L. M. Schiavone, E. Buehler, J. H. Wernick: J. Appl. Phys. **43**, 2805 (1972)

5.99 F. M. Berkovskii, N. A. Goryunova, V. M. Orlov, S. M. Ryvkin, V. I. Sokolova, E. V. Tsvet-kova, G. P. Shpen'kov: Sov. Phys. – Semicond. **2**, 1027 (1969)

5.100 G. P. Shpen'kov: Neorg. Mater. **5**, 1143 (1969)

5.101 B. Sermage, G. Duraffourg: unpublished

5.102 J. L. Shay, B. Tell, H. M. Kasper: Appl. Phys. Lett. **19**, 366 (1971)

5.103 B. Tell, J. L. Shay, H. M. Kasper: Phys. Rev. B**4**, 2463 (1971)
5.104 J. V. Bardeleben, B. Meyer, A. Goltzéné, C. Schwab: Meeting of the French Physical Society, Montpellier, September 5–7, 1973
5.105 J. L. Shay, B. Tell, H. M. Kasper, L. M. Schiavone: Phys. Rev. B**7**, 4485 (1973)
5.106 P. W. Yu: Sol. State Commun. **18**, 395 (1976)

6. Phosphor Films

T. Inoguchi and S. Mito

With 15 Figures

6.1 Background

Intrinsic electroluminescence was discovered by *Destriau* in 1936 [6.1] as an interesting new phenomenon that produces light by exposing phosphor materials to high electric fields, but no positive effort was made in developing the phenomenon into a practical device until engineers at Sylvania Electrical Products, Inc., developed the so-called "powder type" or the "dispersion type" EL panel which was demonstrated at a solid state conference at MIT in the spring of 1952. This drew considerable interest among a number of people, especially those working in illumination engineering and those working on luminescent materials.

One of the reasons why EL was not developed immediately after Destriau's publication may be attributed to the immaturity in the fabrication techniques of transparent electrically conductive films. The transparent conductive film SnO_2 was developed about ten years later.

During the period from 1950 to 1960, basic studies on the dispersion type EL device which consisted of activated phosphor powder dispersed in a layer of insulating substance were carried on very actively with the expectation that it would turn out to be a new light source for wall illumination. Further, it was envisaged that a variety of optoelectronic devices such as light intensifiers, etc., would be developed by the use of EL devices. Although a number of devices were manufactured industrially around 1960, the meagerness in brightness, efficiency and lifetime appeared to be insurmountable. Thus, the first generation of EL development gradually faded away.

In recent years, however, advances in electronics and material science have helped each other in accumulating new knowledge as well as in developing new process technologies and device structures. These seem to have enhanced the development of dispersion type DC-EL in recent years [6.2], but more striking are the advent of the *Lumocen* device [6.3] and the double insulating layer structure [6.4,5] in which the vacuum-deposited thin phosphor film replaces the dispersion layer structure. Problems related to short life and low brightness, the major obstacles that hindered practical application, were both resolved mainly by the latter structure. The vacuum-deposited thin film EL device with a double insulating layer structure not only exhibits high brightness with exceptionally high stability and long life, but it can also be processed to possess inherent memory which lends itself to very useful applications.

The once abandoned, or shelved, EL has now been revived with good prospects of practicability. Among a number of prospective applications, phosphor film EL is looked upon as a promising means for flat information display devices.

In the following section, we would like to describe briefly the development of phosphor film EL devices, mainly from the practical point of view.

6.2 Thin Film EL Devices

6.2.1 LUMOCEN (6.3, 6, 7)

The *Lumocen*, an acronym for "LUminescence from MOlecular CENter", was developed and reported by *Kahng* et al. [6.3] in 1968. As the name LUMOCEN implies, this phosphor has special luminous centers which consist of rare-earth-halide molecules. Figure 6.1 illustrates one of the representative structures of the LUMOCEN device, in which ZnS is normally chosen as the host phosphor material while TbF_3 is the luminous center.

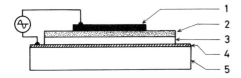

1: Metal Electrode (Al)
2: ZnS:TbF$_3$ (1500 Å)
3: H$_f$O$_2$ (3000 Å)
4: SnO$_2$ (2000 Å)
5: Glass Substrate

Fig. 6.1. Schematic structure of the LUMOCEN

A power conversion efficiency of $\eta = 10^{-4}$ was reported for the ZnS:TbF$_3$ device, which attracted great interest; nevertheless, too little effort was spent in securing operational longevity, so that the device never became a commercial product.

6.2.2 ZnS:Mn Thin Film EL

Studies of Mn-doped ZnS film started rather early in the history of EL. Since 1960, *Vlasenko* et al. have continued their work on ZnS:Mn and have reported several interesting papers [6.8–10]. In 1974, *Hanak* reported on the DC-EL properties in RF sputtered ZnS:Mn$_x$:Cu$_y$ films [6.11]. *Russ* et al. reported in 1967 on a double insulating layer structure thin film EL device [6.4]. These studies seem to have been centered around the clarification of the physics of EL with less emphasis on prolonging the life of operation.

In 1974, the authors reported an extremely long life and high brightness in a device which utilized EL in a thin film of ZnS:Mn with a double insulating layer structure [6.5]. Subsequently, in 1975, inherent memory function was found in the same structure [6.12]. These reports seem to have drawn much attention from EL specialists.

The section to follow deals mainly with the double insulating layer structure ZnS:Mn thin film EL.

6.3 ZnS:Mn Thin Film EL with Double Insulating Layer Structure

6.3.1 Structure and Fabrication of the Devices

Figure 6.2 shows the fundamental structure of the EL device. The device consists of a triple layer structure, namely, an active layer sandwiched between two insulating layers. Because of this sandwich structure, undesirable leakage current flowing through the device is prevented. Consequently, the device can keep a sufficiently high electric field for EL operation across the active layer without breakdown.

Materials and fabrication techniques of these layer are as follows;

i) Transparent Electrode: made of SnO_2 or In_2O_3 using conventional thermal decomposition-oxidation method (SnO_2) and vacuum evaporation method (In_2O_3) on clean glass substrate.

ii) Insulating Layers: High dielectric strength and high dielectric constant material are desirable. Insulating materials such as Y_2O_3, Si_3N_4 and Al_2O_3 are usually satisfactory. Y_2O_3 is easily vacuum evaporated from an electron

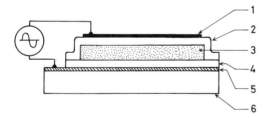

1: Metal Electrode (Al)

2: Insulating Layer ($\doteq$ 2000 Å)

3: Active Layer (ZnS:Mn $\doteq$ 5000 Å)

4: Insulating Layer ($\doteq$ 2000 Å)

5: Transparent Electrode (SnO_2)

6: Glass Substrate

Fig. 6.2. Schematic structure of the thin film EL device

bombarded Y_2O_3 pellet. Si_3N_4 is reactively sputtered from Si cathode, or RF sputtered from Si_3N_4 cathode in $(Ar+N_2)$ atmosphere. Al_2O_3 is vacuum evaporated from an electron bombarded Al_2O_3 pellet, or reactively sputtered from Al cathode in $(Ar+O_2)$ atmosphere. For all these materials, a relatively high deposition rate at low substrate temperature was found effective for realizing excellent dielectric properties.

iii) Active Layer: High-purity ZnS pellet doped with only Mn up to 5 wt.% is used as the source material for the deposition of EL films by means of electron bombardment. During the deposition of ZnS:Mn films, the substrate temperature is maintained around 250 °C and the post deposition annealing is done around 550 °C for one hour in vacuum for stabilization.

iv) Rear Electrode: Usually, formed by vacuum evaporated Al film.

v) Protection Layer: It is necessary to protect the device from humidity, though this is not shown in Fig. 6.2. An outer insulating layer of Si_3N_4 is a satisfactory protection layer.

6.3.2 Performance of the Device

The device fabricated as stated above emits bright yellowish-orange light under alternating electric field excitation. This yellowish-orange light is thought to originate in the Mn center excited directly by hot electrons in ZnS, and has the spectrum shown in Fig. 6.3. This luminescence spectrum does not show any change with driving voltage and frequency.

Figure 6.4 shows the typical brightness-voltage characteristics of the device. The brightness of this thin film EL device is extremely dependent on the applied voltage in the lower voltage region, and tends to saturate in the higher voltage

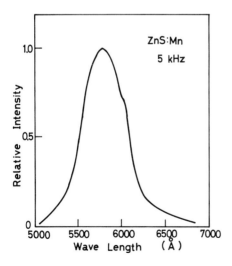

Fig. 6.3. Emission spectrum of ZnS:Mn thin film EL devices

region, as shown in this figure. A typical saturation brightness was about $1.5 \cdot 10^3$ fL at 5 kHz.

In Fig. 6.4, parameters indicated along each curve illustrate the integrated time of operation. As is clear from these curves, the brightness vs. applied voltage characteristics shift to higher voltages as time elapses, but the saturation brightness does not change. This shift of the characteristics was found to settle asymptotically into a final curve after a certain lapse of operation. Consequently, we considered that this shift was not an indication of degradation but an indication of stabilization due to aging effects.

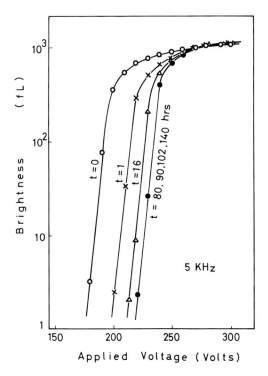

Fig. 6.4. Brightness vs. applied voltage characteristics of the thin film EL device

This stabilizing process was found to be accelerated by elevating the temperature and by applying a suitable voltage on the device at the same time. For instance, if the aging operation was carried out at around 200 °C in the saturation region, this stabilizing process was completed within one hour.

The device became extremely stable after completion of the stabilization process. No change in brightness is observed, as is shown in Fig. 6.5, during an accumulated operation of over $2 \cdot 10^4$ hrs at a constant operating voltage.

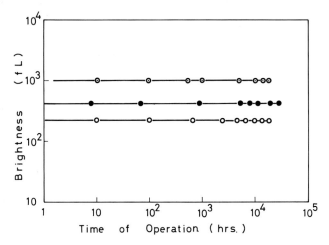

Fig. 6.5. Brightness vs. time of operation at a constant voltage operation. Three samples under different operating conditions are shown

6.3.3 Inherent Memory Function

When the thin film EL device with double insulating layer structure is driven by a series of voltage pulses, the brightness of the device at each pulse is strongly affected by the polarity of the preceding pulse voltage. Fig. 6.6 is a schematic illustration of this phenomenon. As is clear from this figure, when the polarity of the succeeding pulse is inverted, a remarkably high brightness is observed in the device. In contrast to this, if the polarity of the succeeding pulse is the same, the observed brightness is remarkably low. In other words, this thin film EL device can be interpreted as being endowed with memory function. This memory effect was observed to persist for a few minutes under normal room light and for more than 10 hours, when the device was kept in the dark.

The cause of the memory effect is interpreted as follows: The voltage pulse creates a strong electric field which accelerates the electrons within the active ZnS:Mn layer. As they rush across the active layer, these electrons excite the Mn luminous center. Having traveled across the active layer, the electrons accumulate at the interface between the active layer and the insulating layer, beyond which they are not allowed to trespass. These electrons remain at the interface for a rather long period of time even after the electric field is removed (storage effect). A charge polarization across the active layer results. Consequently, if the polarity of the next pulse is in the same direction as that of the first pulse, the effective inner field across the active layer is depressed by the superposition of the previous counter polarization. On the other hand, when the polarity of the next pulse is inverted, the effective innerfield is enhanced by additive polarization (polarization effect). Thus light emission or the output brightness from the device is affected by the polarity of the preceding pulse.

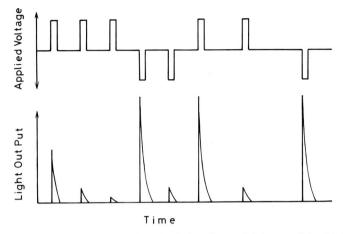

Fig. 6.6. Schematic illustration of polarity effect on brightness of the thin film EL device

Generally, it is considered that the formation of a residual polarization has a remarkable effect on the brigthness of the device driven by a series of voltage pulses depending on its sign and intensity. The residual polarization can be induced not only by electrical means as mentioned above, but also by irradiation with short wavelength light in the presence of a weak electric field (photopolarization effect). This photopolarization effect has a remarkable dependence on the wave length of the irradiating light, as shown in Fig. 6.7.

The above mentioned polarization effects are examples of the inherent memory function. Important practical memory devices can be developed which utilize these polarization effects.

Let us confine ourselves for the time being to memory functions under alternating field excitation.

The brightness vs. voltage-amplitude characteristics of the double insulating layer thin film EL devices manifest a hysteresis-like behavior when the amplitude of applying alternating voltage is increased and then decreased. Fig. 6.8 shows the typical hysteresis behavior that results in the $B-V$ characteristics.

The physics of the phenomenon has not been explained fully as yet. Nevertheless, the foregoing explanation accounts for the qualitative hysteresis phenomenon.

Under excitation with alternating voltage, electrons that are accelerated by the electric field during a half cycle excite the luminous centers. Most of these electrons reach the interface of the insulating layer without being trapped along the way. In the next half cycle, when the electric field is reversed, these electrons add to the electrons accelerated inside the layer and together they excite the luminous centers, thus compensating for the lost electrons. The brightness of the device depends upon the number of electrons accumulated on the interface and is a function of polarizing intensity and it is proportional to the number of electrons that excite the luminous centers.

In the initial stage during which the amplitude of the applied voltage is increased, the mobile electrons that are present in the active layer are those which are thermally excited in the conduction band of ZnS at the ambient temperature. The brightness increases gradually as the amplitude of applied voltage increases in the initial stage. When the amplitude of the increasing applied voltage approaches a threshold value, electrons that had been trapped in certain deep levels inside the triple layer structure begin to contribute to the excitation of luminous center and to the formation of polarization. The rapid increase of brigthness in the $B-V$ characteristics is believed to be due to this liberation of trapped electrons. As long as the amplitude of the applied voltage is kept constant, the electrons liberated from the deep levels continue to contribute to the excitation and polarization.

If the voltage is raised further and all the electrons that had been trapped in the deep levels are released, the number of electrons that contribute to the excitation and polarization becomes constant. This accounts for the saturation behavior of the $B-V$ curve observed at high voltage.

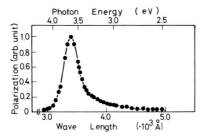

Fig. 6.7. Spectral response of photopolarization effect observed in the thin film EL device

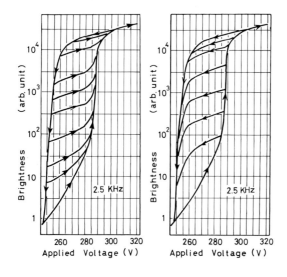

Fig. 6.8. Hysteresis behavior observed in the El brightness vs. voltage characteristics of the double insulating layer structure thin film EL device

In the decreasing stage from the saturation region, the number of electrons is still kept close to the saturated value until the amplitude of the applied voltage decreases to a second threshold value. This keeps the brightness of the device at a higher value than during the increasing stage. Below this second threshold voltage, when the applied voltage is decreased further, the mobile electrons are rapidly trapped by the vacant deep levels, hence the brightness of the device is suddenly depressed until all the empty deep levels are occupied by the electrons. In this manner, in the low voltage region the number of electrons falls again to the value determined by thermal excitation, hence the $B-V$ curve returns to its initial stage.

The existence of the deep level considered here has been verified by observing thermally stimulated current (TSC) across the active layer. Fig. 6.9

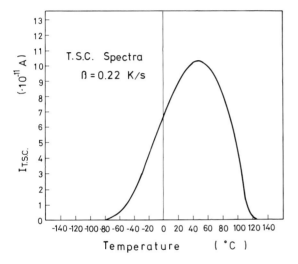

Fig. 6.9. Thermally stimulated current observed in ZnS:Mn layer

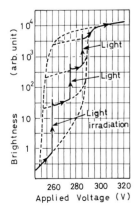

Fig. 6.10. Effect of UV light irradiation on the EL brightness under AC field excitation

shows the TSC observed on ZnS:Mn layer. The depth of these levels has been estimated to be around 0.8 eV from the TSC data.

Irradiation of the EL device by light during alternating voltage excitation gives rise to a change in the brightness through photopolarization and storage effects, as shown in Fig. 6.10. The operating level is considered to be determined by the amount of polarization which in turn depends on the light intensity and the exposure time.

6.4 Thin Film EL as Flat Information Display Panel

As illustrated in Fig. 6.11, the triple layer thin film EL device is readily constructed into cross matrix type displays. An attempt to realize EL flat information displays had been carried out in the first generation of EL displays as was mentioned in Section 6.1.

Efforts were spent in developing flat information display panels using phosphor powders in dispersion type EL panels. This development was particularly aimed at obtaining gray scale picture displays [6.13,14]. Although this effort persisted up to recent years, the brightness was insufficient for practical use and the scattering of light by the powder particles degraded the contrast. Furthermore, the poor stability of operation prevented the display from evolving into immediate applications.

The DC excited powder type EL panel seems to have been developed successfully into cross matrix panels [6.2,15].

The thin film EL device, on the other hand, consists of a few layers of transparent films and a single metal electrode, thus getting away from the degradation of contrast in the dispersion type EL caused by the rather coarse phosphor particles. Besides, the thin film EL provides high brightness and excellent stability and it is expected to become a practical device.

The thin film EL (TF-EL) when applied to flat information display panels compares favorably with other flat panels such as plasma display panels (PDP) or liquid crystal display panels (LCD) as follows:

1) TF-EL is an all solid state device unlike the PDP or LCD. 2) TF-EL offers faster response and wider viewing angle than those of LCD. 3) The

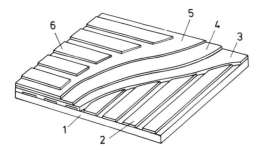

Fig. 6.11. Schematic construction of the thin film EL matrix type information display panel. **1** Glass substrate, **2** Transparent electrodes, **3** Insulating layer, **4** Active layer, **5** Insulating layer, **6** Rear electrodes

TF-EL is not only capable of ON-OFF binary switching but is also capable of recording information with gray scale electrically and optically making use of the inherent memory function. This promises the development of a multi-function display device capable of a variety of applications.

6.4.1 Thin Film EL TV Display Panel

The display of TV pictures was demonstrated experimentally with a matrix type thin film EL panel without memory function [6.16]. The panel had a display area of $48 \cdot 36$ mm^2, consisting of 120 vertical and 90 horizontal electrodes and a triple layer construction. Gray scale was realized by pulse

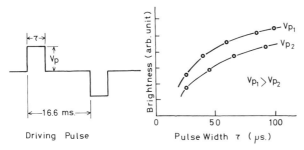

Fig. 6.12. Brightness vs. driving pulse width characteristics of the thin film EL device

Fig. 6.13. Output image produced on matrix-type thin film EL panel having $90 \cdot 120$ electrodes in $36 \cdot 48$ mm^2 area

width modulation as illustrated in Fig. 6.12, in which a 3-bit, 8 grade display was adopted. A standard broadcast TV signal was displayed by sampling one out of three horizontal lines in a line sequential system. The highlight brightness was 60 fL and the maximum contrast ratio was 50 to 1. Although sampling was adopted, no cross talk was observed (see Fig. 6.13).

6.4.2 Application of Memory Function of Thin Film EL Panels to Information Displays

In the double insulating layer thin film EL device a hysteresis behavior is present in the characteristics of brightness versus AC voltage (see Fig. 6.8). If this hysteresis effect is utilized, the display panel can be endowed with a memory function, which is indispensable to the matrix type information display panel.

Figure 6.14 describes the memory function that is realized by utilizing this effect.

V_s denotes the amplitude of the sustaining voltage. The amplitude of the sustaining voltage is normally adjusted to be in the vicinity of the central portion of the hysteresis region. Initially, the brightness is B_0. Synchronized with this sustaining voltage, a writing pulse of the same polarity is superposed, whose resultant voltage is denoted by V_w. The brightness rises instantly to B_w'. The voltage is then set back to its initial sustaining value V_s, but the bright-

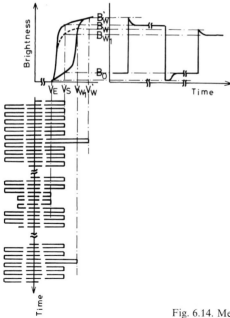

Fig. 6.14. Memory operation based on *B-V* hysteresis loop

ness stays at B_w due to the hysteresis, indicating the completion of writing in. The written-in brightness B_w is held as long as the sustaining pulse voltage V_s remains the same. Erasing is accomplished by superposing a counterpolarized pulse upon the sustain pulse, the resultant of which is V_E. After a few successive erasing pulses, the voltage is returned to the initial sustaining voltage V_s, whereupon the brightness returns to B_0 again. In this manner, and due to the presence of minor hysteresis loops, the written-in brightness may be selected at any value between B_w and B_0. Hence, a gray scale may be realized by using the memory effect, instead of pulse-width modulation.

6.4.3 Thin Film EL Character Display Panel with Inherent Memory

A double insulating layer thin film EL device endowed with inherent memory was constructed as a cross matrix panel with $240 \cdot 320$ dot elements in an area of $120 \cdot 160$ mm² [6.12]. As described in the preceding section, this panel is capable of displaying gray scale pictures. An alpha-numeric character display that operates in a non-refreshing mode was constructed and tested [6.17].

The experimental results are briefly summarized as follows,
1) Number of displayed characters: 1248 characters (52 ch. $\cdot$ 24 lines).
2) Color of display: Yellowish orange.
3) Brightness: About 30 fL at 300 Hz.
4) Contrast ratio: 10 to 1.
An example of the displayed characters is shown in Fig. 6.15.

Fig. 6.15. Character display on matrix type thin film EL panel having $240 \cdot 360$ electrodes in $120 \cdot 160$ mm² area

6.5 Conclusion

The development of intrinsic EL by means of phosphor film stagnated in the first half of 1960s, and most of the applications were taken over by LEDs. However, in recent years, the advent of the AC type thin film EL has indicated a bright prospect for practical application to various kinds of displays due to its high reliability, brightness, life, and its inherent memory function.

The DC-EL performance of powder type and thin film type structures has been improved several times.

In a few years, it may be expected that the AC type thin film EL will become practical for character displays. It is believed that this panel is competitive with PDP in brightness, reliability, and in memory function. The all solid state flat display panel has a promising viability in the field of information displays.

The final application would be in the flat panel TV display. Problems such as ortho-chromatic display, conversion efficiency, and cost should be resolved before its penetration into the consumer market. The former two will be solved by further studies of the materials. As for the cost, the general trend of its reduction will be in accord with the experience curve that will be affected by the improvement in mass production technique of the panel and the number of units marketed.

References

6.1   G. Destriau: J. Chem. Phys. **33**, 620 (1936)
6.2   A. Vecht: J. Vac. Sci. Tech. **10**, 789 (1973)
6.3   D. Kahng: Appl. Phys. Lett. **13**, 210 (1968)
6.4   M. J. Russ, D. I. Kennedy: J. Electrochem. Soc. **114**, 1066 (1967)
6.5   T. Inoguchi, M. Takeda, Y. Kakihara, Y. Nakata, M. Yoshida: '74 SID Intern. Symp. Digest (1974) pp. 84–85
6.6   E. W. Chase, R. T. Hepplewhite, D. C. Krupka, D. Kahng: J. Appl. Phys. **40**, 2512 (1969)
6.7   Y. S. Chen, M. V. DePaolis, Jr., D. Kahng: Proc. IEEE **58**, 184 (1970)
6.8   N. A. Vlasenko: Opt. Spect. **18**, 260 (1965)
6.9   N. A. Vlasenko, A. M. Yaremko: Opt. Spect. **18**, 263 (1965)
6.10 N. A. Vlasenko, S. A. Zynio, Yu. V. Kopytko: phys. stat. sol. (a) **29**, 671 (1975)
6.11 J. J. Hanak: Proc. 6th Intern. Vacuum Cong., Kyoto (1974) pp. 809–812
6.12 M. Takeda, Y. Kakihara, M. Yoshida, M. Kawaguchi, H. Kishishita, Y. Yamauchi, T. Inoguchi, S. Mito: '75 SID Intern. Symp., Sect. 7.8 (1975)
6.13 H. Arai, T. Yoshizawa, K. Awazu, K. Kurahashi, S. Ibuki: Proc. 1970 IEEE Conf. Display Devices (New York 1970) p. 52
6.14 M. Yoshiyama, H. Kawarada, T. Sato: Proc. IEEE Intern. Computer Group Conf. (Wash., D.C., 1970) pp. 261–269
6.15 A. Vecht, N. J. Werring, R. Ellis, P. J. F. Smith: Proc. IEEE **61**, 902 (1973)
6.16 S. Mito, C. Suzuki, Y. Kanatani, M. Ise: '74 SID Intern. Symp. Digest (1974) pp. 86–87
6.17 C. Suzuki, Y. Kanatani, M. Ise, E. Mizukami, K. Inazaki, S. Mito: '76 SID Intern. Symp. Digest (1976) pp. 50–51

Subject Index

Applied Physics

A monthly journal

Board of Editors **S. Amelinckx,** Mol. · **V. P. Chebotayev,** Novosibirsk
R. Gomer, Chicago, Ill. · **H. Ibach,** Jülich
V. S. Letokhov, Moskau · **H. K. V. Lotsch,** Heidelberg
H. J. Queisser, Stuttgart · **F. P. Schäfer,** Göttingen
A. Seeger, Stuttgart · **K. Shimoda,** Tokyo
T. Tamir, Brooklyn, N.Y. · **W. T. Welford,** London
H. P. J. Wijn, Eindhoven

Coverage application-oriented experimental and theoretical physics:

Solid-State Physics   *Quantum Electronics*
Surface Physics   *Laser Spectroscopy*
Chemisorption   *Photophysical Chemistry*
Microwave Acoustics   *Optical Physics*
Electrophysics   *Integrated Optics*

Special Features **rapid** publication (3–4 months)
no page charge for **concise** reports
prepublication of titles and abstracts
microfiche edition available as well

Languages Mostly English

Articles original reports, and short communications
review and/or tutorial papers

Manuscripts to Springer-Verlag (Attn. H. Lotsch), P.O. Box 105 280
D-69 Heidelberg 1, F.R. Germany

Place North-American orders with:
Springer-Verlag New York Inc., 175 Fifth Avenue, New York. N.Y. 10010, USA

Springer-Verlag
Berlin Heidelberg New York

Springer Series in Optical Sciences

Editor: D. L. MacAdam

Springer-Verlag Berlin Heidelberg New York